TECHNIQUES IN DNA SEQUENCING

TECHNIQUES IN DNA SEQUENCING

Dr. R.N. Shukla

Oxford Book Company

Jaipur ▪ New Delhi

Published By

Oxford Book Company

Regd. Off. "Bony Residency", Gate No. 2
Opp. Tilak Public School, Vishweswariya Nagar,
Gopalpura Road, Jaipur - 302018, India
Ph: 0141 - 2761280, Telefax: 2761381
email - oxfordbookjpr@gmail.com
website - oxfordbookcompany.com

Delhi Office: 102, 1st Floor, Satyam House,
4327/3, Ansari Road, Darya Ganj,
New Delhi - 110002, India
Ph: 011-45652440

Distributed By

KEYA INTERNATIONAL PRESS LLP
E-mail: keyainternationalpressllp@gmail.com

Published in India, 2025

ISBN : 9789391096564

Typeset by Shivangi Computers, Jaipur

Printed and bound in India by Thomson Press Ltd.,
New Delhi

Contents

Preface

The cells, instead of making more cells by division, grow bigger and biggerwithout division. Their nuclei, too, grow bigger and bigger. Inside the nuclei, the chromosomes make replicas of themselves as they doin preparation for mitosis but, as there is no mitosis, the new chromosomes do not separate from the old ones. Instead, new and old chromosomes remain closely attached to each otherand each, in turn, makes more replicas. This goes on repeatedly, and so accurately does each band make another band like itself that the whole bundle appears as one single thick banded thread. In addition, the partner cromosomes in the salivary glands are closely paired as in meiosis, so that each apparent chromosome represents one pair, and thenuclei of the salivary glands of Drosophila seem to have four chromosomes instead of the eight that are visible in ordinary cells. The pattern of bands corresponds to the pattern of underlying genes. This can often be seen in cases where a chromosome has lost a small piece through radiation or some other injury. The accuracy of chromosome replication is such that all descendants of the damaged chromosome lack exactl the same piece, and when such a damaged chromosome gets into a gamete, for instance an ovum, individuals are formed which lack the same chromosome piece in all their cells. The left part of a normal wing of Drosophila and above it, and at a very much higher mgnification, a little section of one of the normal salivary gland chromosomes. The arrow points to a band which encloses a gene for wing shape. The right part of the same chromosome section in afly which had received a damaged chromosome from its X-rayed father or mother.

If you recall that the salivary gland chromosomes consist of two chromosome bundles formed by the two partner chromosomes, you will unerstand why only half a band is missing at the point of the arrow. Nevertheless, the lack of one of the two partner genes has been sufficient to disturb normal development of thewing and to produce the abnormal shape shown in the accompanying sketch. Flies which have lost this band from both partner chromosomes are so much damaged thatthey die as early embryos. The left drawing represents a normally growing green maize plant and underneath, at much higher magnification, the tips of one particular chromosome pair. Thee is a fairly large knob at the end of these chromosomes. Among the genes which this knob encloses is one whose action is necessary for the production of the green substance chlorophyll that is essential for survival of the plant.

The right picture shows that, when this knob has been lost rom both chromosomes, the plant dies early as a yellow and sickly seedling. If you turn once more to the left picture you will notice a striking difference from what happened in the case of the Drosophila wing gene. Whereas this gene could not produce anormal wing without the co-operation of its partner gen, the chlorophyll gene is sufficiently effective even when acting singly; for plants with one normal and one knobless chromosome are quite normal. Tis is not a distinction between the genes of Drosophila and maize, or between genes concerned with wing hape or chlorophyll. It is a difference in the efficiency with which genes carry out their functions; some genes are elfsufficient in this regard, others need the co-operation of their partners for effective action. We shall meet a similar difference when we deal with dominant and recessive genes. Each gene is concerned in the control of a particular developmetal process, for instance the formation of chlorophyll, the development of colour or size or, in higher animals, mental ability. Partner genes control the same process, but they may do so in different ways; thus two partner genes in peas are concerned with the control of seed shape, but one causesthe seeds to be round, while the other makes them wrinkled.

This book is a comprehensive and analytical study of Molecular Genetics which covers, nearly, all important aspects of DNA and molecular testing.

— Dr. R.N. Shukla

Chapter 1

The Cell

The cell is one of the most basic units of life. There are millions of different types of cells. There are cells that are organisms onto themselves, such as microscopic amoeba and bacteria cells. And there are cells that only function when part of a larger organism, such as the cells that make up your body. The cell is the smallest unit of life in our bodies. In the body, there are brain cells, skin cells, liver cells, stomach cells, and the list goes on.

All of these cells have unique functions and features. And all have some recognizable similarities. All cells have a 'skin', called the plasma membrane, protecting it from the outside environment. The cell membrane regulates the movement of water, nutrients and wastes into and out of the cell. Inside of the cell membrane are the working parts of the cell. At the centre of the cell is the cell nucleus.

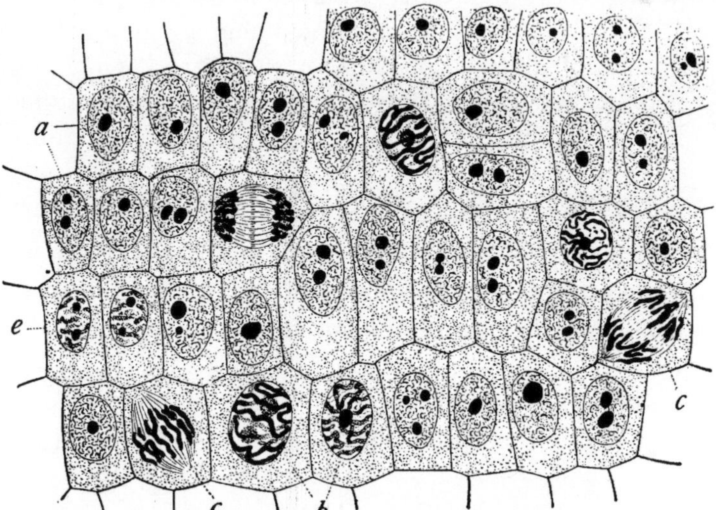

Figure 1: Allium cells in different phases of the cell cycle

The cell nucleus contains the cell's DNA, the genetic code that coordinates protein synthesis. In addition to the nucleus, there are many organelles inside of the cell-small structures that help carry out the day-to-day operations of the cell. One important cellular organelle is the ribosome. Ribosomes participate in protein synthesis.

The transcription phase of protein synthesis takes places in the cell nucleus. After this step is complete, the mRNA leaves the nucleus and travels to the cell's ribosomes, where translation occurs. Another important cellular organelle is the mitochondrion. Mitochondria (many mitochondrion) are often referred to as the power plants of the cell because many of the reactions that produce energy take place in mitochondria. Also important in the life of a cell are the lysosomes. Lysosomes are organelles that contain enzymes that aid in the digestion of nutrient molecules and other materials. Below is a labelled diagram of a cell to help you identify some of these structures.

There are many different types of cells. One major difference in cells occurs between plant cells and animal cells. While both plant and animal cells contain the structures, plant cells have some additional specialized structures. Many animals have skeletons to give their body structure and support. Plants do not have a skeleton for support and yet plants don't just flop over in a big spongy mess. This is because of a unique cellular structure called the cell wall.

The cell wall is a rigid structure outside of the cell membrane composed mainly of the polysaccharide cellulose. As pictured at left, the cell wall gives the plant cell a defined shape which helps support individual parts of plants. In addition to the cell wall, plant cells contain an organelle called the chloroplast. The chloroplast allow plants to harvest energy from sunlight. Specialized pigments in the chloroplast (including the common green pigment chlorophyll) absorb sunlight and use this energy to complete the chemical reaction:

$$6\ CO_2 + 6\ H_2O + \text{energy (from sunlight)} \rightarrow C_6H_{12}O_6 + 6\ O_2$$

In this way, plant cells manufacture glucose and other carbohydrates that they can store for later use.

Organisms contain many different types of cells that perform many different functions. In the next lesson, we will examine how individual cells come together to form larger structures in the human body.

Each cell has a part called a cell membrane; a thin covering around the cell. The cell membrane separates it, gives the cell its shape, and controls what goes into and out of it. Located inside a cell is the cytoplasm. Cytoplasm is a jellylike liquid. Inside the cytoplasm is the cell's control centre called the nucleus. The nucleus contains all of the instructions for running the cell. The nucleus is surrounded by the nuclear membrane. The nuclear membrane controls what goes into and comes out of the nucleus.

A cell is able to reproduce by means of splitting. One cell splits into two cells. However before the cell splits, the nucleus doubles and divides in half so each new cell has a copy of the cell's instructions.

Look carefully at the diagram of the cell below. Notice inside the cytoplasm are vacuoles and mitochondria. Vacuoles are tiny oval structures that store food, water, or wastes. Mitochondria are shaped like kidney beans. Mitochondria are important parts of a cell because they help change food into energy. The cell uses the energy to do its work.

Cells do not all look the same. Their shape depends upon the jobs they perform. Red blood cells are tiny and round. They need to be so they can squeeze through blood vessels and bring oxygen to the other cells in the body. Muscle cells are thin and long. This is so they can expand and contract to help the body move.

All cells, whether plant or animal have these same parts: cell membrane, cytoplasm, nucleus, nuclear membrane, vacuoles, and mitochondria. Plant cells however have two additional parts that animal cells do not. Plant cells have cell walls. This sturdy layer around the cell membrane supports and protects the cell. Plant cells also contain chloroplasts. Chloroplasts contain the green substance called chlorophyll. Chlorophyll traps energy from sunlight and enables a plant to make its own food.

The Cell Cycle

All organisms are composed of cells. Organisms may be either single celled, as found in the Kingdoms of Archaebacteria and Eubacteria and Protista; or multicellular, as found in the Kingdoms Fungi, Plantae and Animalia. While all of the organisms found on Earth are quite diverse, they all have several characteristics in common:

- the ability to metabolize, or maintain the biochemical processes necessary for life
- the ability to grow and evolve
- the ability to reproduce.

All of the organisms in the Kingdom Archaebacteria and Eubacteria are prokaryotic bacteria. These single celled organisms have no nucleus and no complex membrane bound organelles. They reproduce by an asexual process called binary fission. Binary fission means literally "splitting into two," and involves no genetic recombination. Thus, this process results in the formation of identical cells, or clones. Because this is true for all bacteria you may wonder how different strains of bacteria arise. For example, how does non-pathogenic *Escherichia coli* (*E. coli*), identical to that found in your gut somehow produce killer *E. coli* strain O157:H7, which caused many deaths in the western United States in 1993 and continues to pose a threat? The many variations within a given species of bacteria arise from mutation of its DNA by either mutagenic agents such as UV light or by viral infection of the bacterial cell.

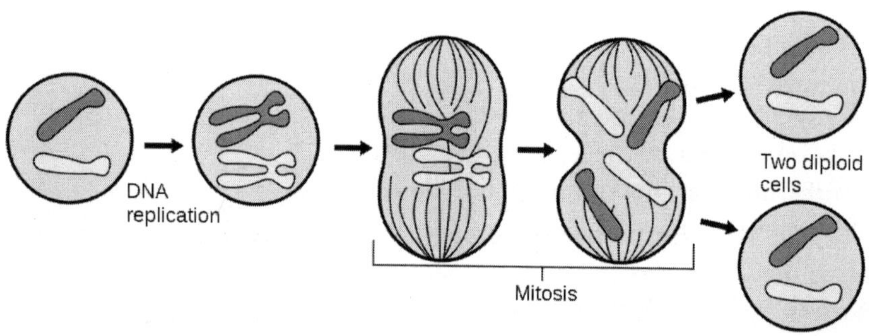

Figure 2: Each turn of the cell cycle divides the chromosomes in a cell nucleus.

While mutagenic agents and viruses can also alter eukaryotic cells, the effect is generally not seen in the offspring. The primary force contributing to genetic diversity in the multicellular eukaryotes is sexual reproduction. Sexual reproduction involves the recombination of genetic material between two different individuals of the same species to produce a genetically unique offspring. For example, you obtained one half of your genetic information (DNA) from your mother and one half from your father-thus you are a combination of your two parents and quite genetically distinct from all other humans (unless you have an identical twin). You came into being when two haploid gametes (sperm and egg) fused and became a single diploid cell, called a zygote.

Since you started as a single cell, how did you become the complex multicellular organism you are today? The zygote's first job is to divide and produce many identical cells, through the processes of mitosis and cytokinesis. Mitosis involves the division of the nucleus, and cytokinesis involves the division of the cytoplasm. This results in the formation of two identical cells from one cell.

How does mitosis and cytokinesis occur, and what does the cell do between these division events? Both of these processes occur only after cell growth, metabolism, and DNA duplication within the nucleus of the cell. Cells within a multicellular organism have a distinct life span, which is called the cell cycle. Consider the cell cycle as the life span of a single cell.

The cell is "born" after cell division, it grows and metabolizes in a cell cycle phase called G1 phase, and then it prepares to reproduce. First, it must duplicate all of its genetic information, or DNA, in a cell cycle phase called S phase. Then it must spend some time metabolizing all the necessary materials that it needs to split into two cells during the phase called G2 phase. Finally the nucleus of the cell, which now has double the amount of its DNA, goes through nuclear division. This nuclear division is called mitosis. When mitosis is almost complete, the cell cytoplasm must split. The formation of new cell wall or membrane then results in two identical cells.

Each type of cell, whether epithelial, cardiac, nerve, or some other type, has its own distinct cell cycle, so the time spent in each phase may vary between different cell types. However, all eukaryotic cells in multicellular organisms have the same generalized cell cycle containing G1, S, G2, Mitosis, and Cytokinesis phases.

Homologous Chromosomes, Genes, and DNA

Almost everyone in the media talks about chromosomes, DNA, and genes as if they were all the same entity. While they are related, one must realize the relationship between these terms to gain a real understanding of how the most basic of life processes occur. Chromosomes are highly condensed (tightly wound), huge biomolecules, composed of DNA, protein, and some RNA. Some organisms contain only one unpaired chromosome that is not contained in a nuclear envelope. All of these organisms are called prokaryotes.

Eukaryotes, by contrast, especially in multicellular organisms, have several chromosomes per cell. These chromosomes exist as paired, or homologous, chromosomes. The homologous chromosomes condense and pair during the mitosis stage of the life cycle of all eukaryotes.

The reason these chromosomes are paired is that half of an organism's chromosomes came from the paternal gamete and the other half came from the maternal gamete.

Let's use human beings again as an example. Humans have 46 individual chromosomes in every cell of their body during mitosis, except for the gametes (sex cells like eggs or sperm) and mature red blood cells. However, there are only 24 different human chromosomes. How can this be?

Humans acquire a set of 22 chromosomes and one X chromosome from their mother for a total of 23, and a set of 22 chromosomes and one X or Y chromosome from their father to make up the 46 chromosomes. Does the mother's chromosome number five differ from the father's chromosome number five? Yes! The two chromosomes do differ in their specific information. While both homologous chromosomes contain the same type of information located at the same exact place, they each have a different version of the information.

The chromosome from the one parent may contain information for blue eye colour, and the chromosome from the other parent may contain information for brown eye colour. While the specific DNA sequence information may vary from individual to individual, chromosome number five in humans encodes for the same type of product.

The pairs of homologous chromosomes contain redundant information so that you have two copies of each chromosome, one from Mom and one from Dad. In this way, you may exhibit traits from both parents.

As you probably know, human males have two different sex-determining chromosomes: one X and one Y. These are not homologous, but contain very different information. Females humans have two X chromosomes. So besides the regular 22 chromosomes, there are also two different sex chromosomes, for a total of 24 different chromosomes in humans.

How exactly do genes relate to chromosomes? First you must realize that DNA, the essential molecule of the chromosome, contains all the information of the cell for the manufacture of all proteins necessary to maintain and perpetuate life. A gene is a segment of DNA that encodes for one polypeptide, or molecule of protein. The colour of your eyes results from the production of specific polypeptides, which come together to form a protein that exhibits physical properties that makes them appear to have a certain colour. These polypeptides are coded by genes or a set of genes. The genes are DNA sequences contained in a chromosome.

There are many genes on a single chromosome. Note that much of the DNA in a chromosome does not encode for a polypeptide and is not gene DNA. The function of non-genic DNA is not well understood, although some parts of this DNA are known to regulate the expression of genes.

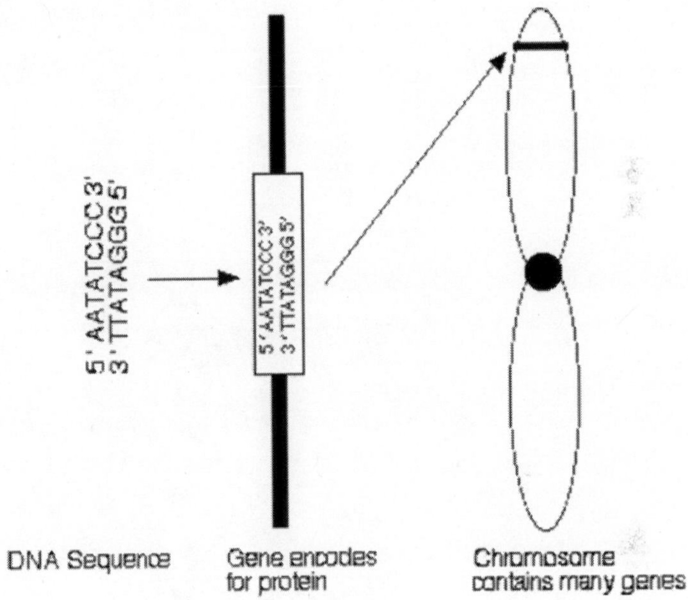

Figure 3: The relationship between DNA, genes, and chromosomes.

Mitosis: Division of the Cell Nucleus

Now to discuss how the process of mitosis occurs in a cell. It is very important to remember that the cell must have duplicated all of its DNA in S phase of the cell cycle, before mitosis can occur.

If an organism has three pairs of homologous chromosomes, how many individual chromosomes does it have? How many chromosomes does this same cell have after S phase of the cell cycle?

Let's discuss the terms diploid and haploid. Any organism or cell that has paired, homologous chromosomes is diploid. If a cell or organism has only one chromosome of each type it is considered haploid. Ploidy does not deal with the numbers of chromosomes, but instead deals with having paired or unpaired chromosomes. Most cells in your body are diploid and contain two of each type of chromosome.

Gametes, or sex cells, only contain one set of chromosomes, so they contain half the diploid number. Gametes are considered haploid cells, having only one of each of the homologous chromosomes. Now to answer the above questions: an organism having three pairs of homologous chromosomes has six individual chromosomes. After S phase when the DNA replicates, the cell contains 12 individual chromosomes. Now the cell can form two nuclei, each having one pair of each type of chromosome, and the cell can split to form two identical cells.

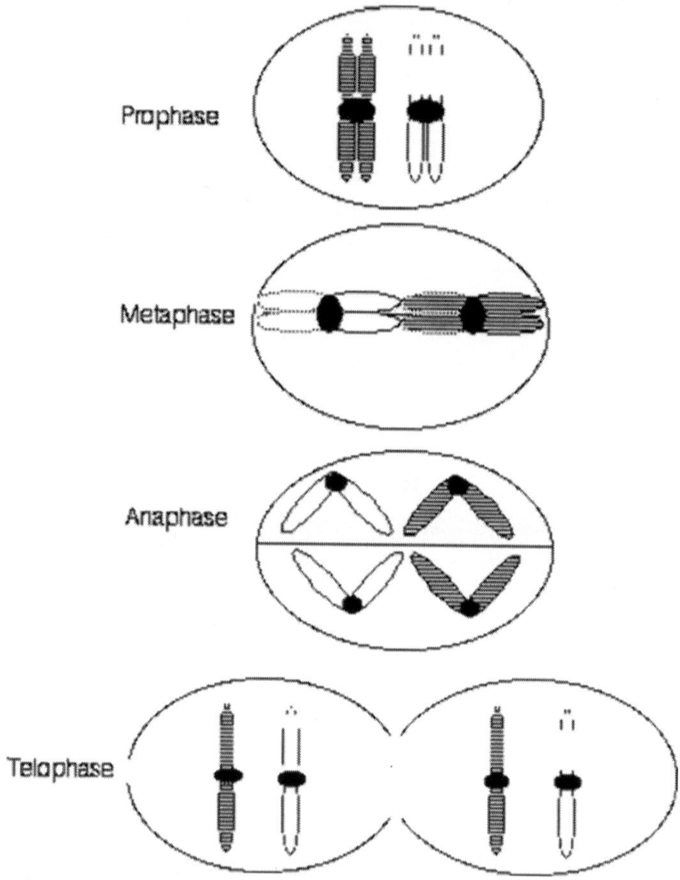

Figure 4: Representation of a cell going through mitosis.

Mitosis occurs in several stages. Remember that these stages are part of a very rapid, continuous process and are not discrete events as might seem from their descriptions. It is important to learn the names of each stage and to be able to describe the cellular processes that occur.

Outline of stages of Mitosis and Cytokinesis:

I. Mitosis:

a. Prophase

b. Metaphase

c. Anaphase

d. Telophase

II. Cytokinesis.

The most important information to learn about Mitosis is the position and number of chromosomes in each of the stages of Mitosis. You must be able to answer a question similar to the following example: diagram a cell that has 2 pairs of homologous chromosomes in each of the stages of mitosis.

Prophase: The first stage of Mitosis when chromosomes condense and the nuclear membrane disappears. The chromosomes that have duplicated in S phase attach to their duplicate in pairs at a central point called a centromere. At this stage the attached chromosomes are called sister chromatids. Sister chromatids are identical and are very different from homologous chromosomes. All these sisters are identical twins. Each homologous chromosome has been duplicated, and now the two duplicates are attached to one another to form sister chromatids.

Metaphase: The stage at which the chromosomes line up at the equator of the cell and spindle fibres attach to the centromere of each chromosome. The spindle fibres, made of protein subunits, shorten and start to pull the chromosomes to opposite sides of the cell.

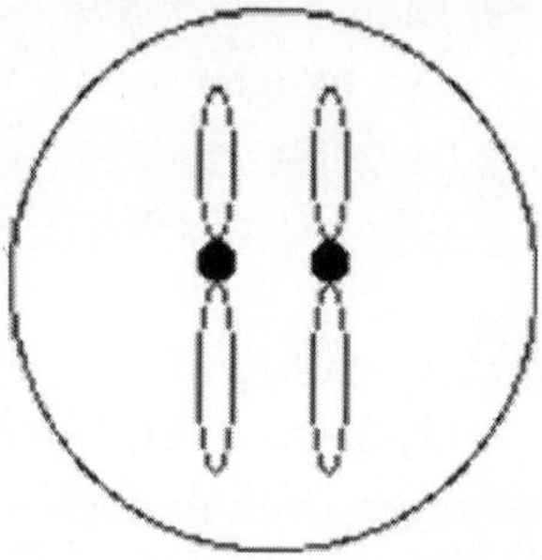

Figure 5: Diploid cell containing one pair of homologous chromosomes before S phase of the cell cycle. This is what a normal diploid cell would look like if chromosomes were condensed before S phase occurred (they aren't actually condensed until prophase) and what both cells should look like immediately following telophase, before the chromosomes decondense.

Anaphase: The stage at which sister chromatids (identical duplicates) are separated and pulled to opposite poles of the cell.

Telophase: The stage at which the cell membrane or cell wall forms in the centre of the cell.

Cytokinesis: The stage where the cytoplasm and organelles are divided between the two cells and the chromosomes decondense, forming two diploid, identical cells.

Meiosis: the Formation of Gametes

We have previously discussed the concept of sexual reproduction. Generally, in most organisms the chromosome content of the cell must be diploid, or 2N, where N represents the haploid number of chromosomes. In other words, cells usually do not tolerate more than two homologous chromosomes in the cell at one time. For example, if a human zygote were formed by the fusion of two diploid cells, rather than two haploid gamete cells, the zygote would have 92 chromosomes. That translates to four sets of the 22 chromosomes and four sex chromosomes. You can see that this would soon become a huge problem for the cell!

To avoid these problems, most organisms donate only one half of a set of chromosomes during sexual recombination. The resulting zygote will therefore have exactly the same number of chromosomes as the parents. The process by which the cell accomplishes this feat is called Meiosis. Meiosis is a type of cell division that occurs only in diploid cells destined to become gametes. No other type of cell undergoes Meiosis.

Before Meiosis begins, the diploid cell goes through the entire cell cycle including S phase, just exactly as a cell does before Mitosis occurs. Since the cell now has double the number of chromosomes of the original diploid cell, it must go through two rounds of division to become haploid.

Example: A cell that contains two pairs of homologous chromosomes, or four chromosomes in all, goes through S phase before Meiosis and now has four pairs of homologous chromosomes, or eight chromosomes in all. The first meiotic division results in two cells each having two pairs of homologous chromosomes. The second meiotic division, undergone by both cells, results in four cells each having two chromosomes, one of each homologous pair.

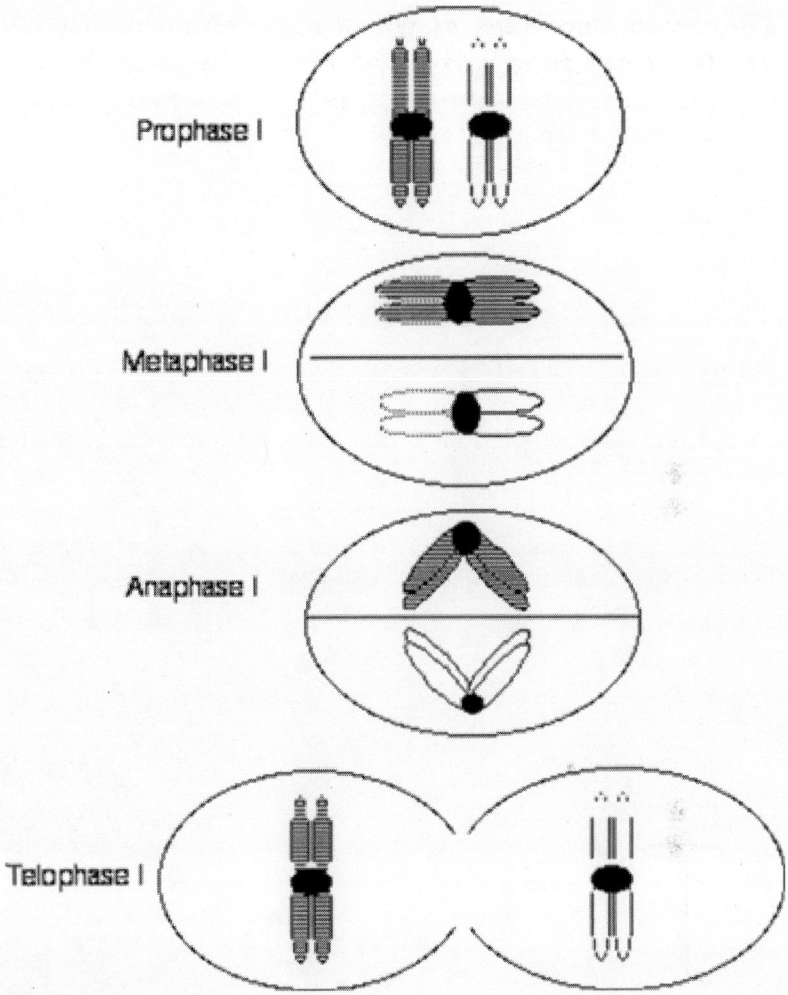

Figure 6: Meiosis I.

The process of Meiosis has many similarities to Mitosis. Again, you should write a brief description for each of the following phases of Mitosis.

I. Meiosis I

a. Prophase I

b. Metaphase I

c. Anaphase I

d. Telophase I

II. Cytokinesis

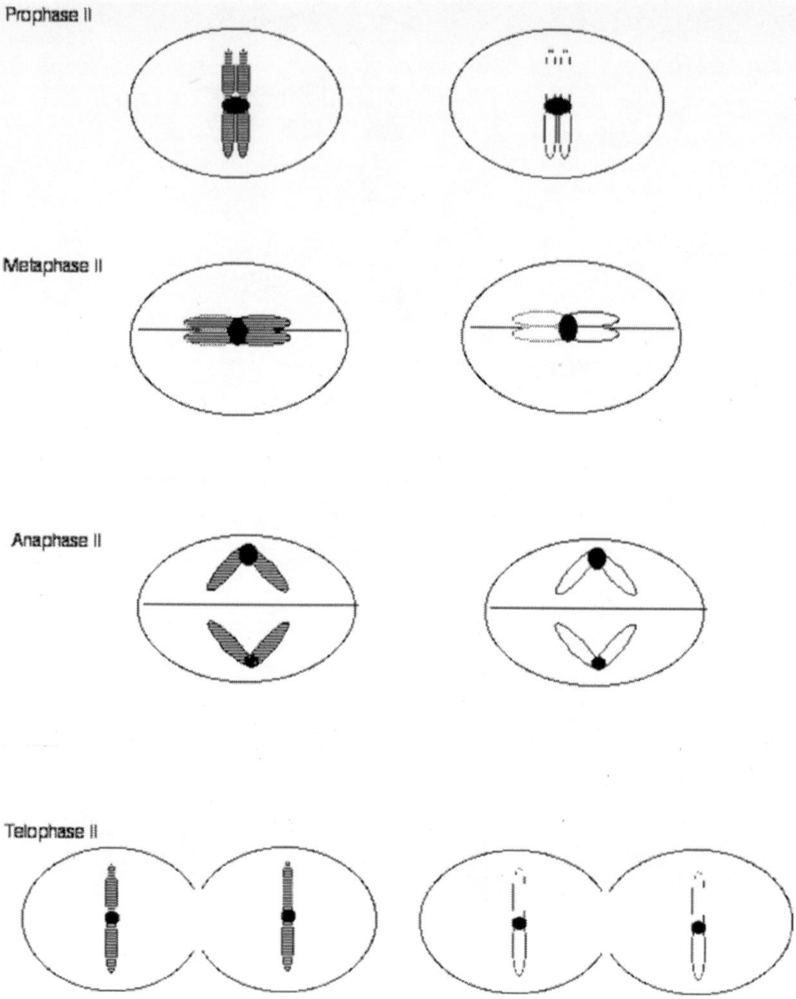

Figure 7: Meiosis II.

III. Meiosis II
   a. Prophase II
   b. Metaphase II
   c. Anaphase II
   d. Telophase II

IV. Cytokinesis

The most important information for you to know about Meiosis is the position and number of chromosomes in each of the stages. You must be able to answer a question similar to the following example:

Please diagram a cell which has two pairs of homologous chromosomes in each of the stages of Meiosis.

Prophase I: The first stage of Meiosis when chromosomes condense and the nuclear membrane disappears. The chromosomes that have duplicated in S phase attach to their duplicate in pairs at a central point called a centromere. At this stage the attached chromosomes are called sister chromatids. Sister chromatids are identical and are very different from homologous chromosomes. All these sisters are identical twins. Each homologous chromosome has been duplicated, and now the two duplicates are attached to one another to form sister chromatids.

Metaphase I: The stage at which the chromosomes line up at the equator of the cell and spindle fibres attach to the centromere of each chromosome. The spindle fibres, made of protein subunits, shorten and start to pull the chromosomes to opposite sides of the cell.

Anaphase I: The stage at which sister chromatids (identical duplicates) stay together and are pulled to opposite poles of the cell.

Telophase I: The stage at which the cell membrane or cell wall forms in the centre of the cell.

Cytokinesis: The stage where the cytoplasm and organelles are divided between the two cells and the chromosomes decondense, forming two diploid, identical cells.

And now for Meiosis II...

Prophase II: The stage of Meiosis II when chromosomes condense and the nuclear membrane disappears. The chromosomes have not been duplicated, and the cell is still only diploid.

Metaphase II: The stage at which the chromosomes line up at the equator of the cell and spindle fibres attach to the centromere of each attached sister chromatid.

Anaphase II: The stage at which sister chromatids separate and are pulled to opposite poles of the cell.

Telophase II: The stage at which the cell membrane or cell wall forms in the centre of the cell.

Cytokinesis: The stage where the cytoplasm and organelles are divided between the two cells and the chromosomes decondense, forming two haploid gamete cells. The gamete has one half of the information of a diploid cell. Each of the four gametes produced has one of each homologous chromosome, in contrast to a diploid cell that has homologous pairs of chromosomes.

Enzymes: The Catalysts of Life

Each cell in the body is a beehive of activity called metabolism. The biochemical activity is the process of chemical and physical change which goes on continually in the living organism. These characteristics of life include the build-up of new tissue, replacement of old tissue, conversion of food to energy, disposal of waste materials, and reproduction.

This building up and tearing down takes place in the face of an apparent paradox. The greatest majority of these biochemical reactions do not take place spontaneously. The phenomenon of catalysis makes possible biochemical reactions necessary for all life processes. Catalysis is defined as the acceleration of a chemical reaction by some substance which itself undergoes no permanent chemical change. The catalysts of biochemical reactions are enzymes and are responsible for bringing about almost all of the chemical reactions in living organisms. Without enzymes, these reactions take place at a rate far too slow for the pace of metabolism.

The oxidation of a fatty acid to carbon dioxide and water is not a gentle process in a test tube – extremes of pH, high temperatures and corrosive chemicals are required. Yet in the body, such a reaction takes place smoothly and rapidly within a narrow range of pH and temperature. In the laboratory, the average protein must be boiled for about 24 hours in a 20% HCl solution to achieve a complete breakdown. In the body, the breakdown takes place in four hours or less under conditions of mild physiological temperature and pH.

It is through attempts at understanding more about enzyme catalysts – what they are, what they do, and how they do it– that many advances in medicine and the life sciences have been brought about.

All known enzymes are proteins. They are high molecular weight compounds made up principally of chains of amino acids linked together by peptide bonds.

The primary benefit of enzymes for us is in digestion...

Digestive Enzymes

Specific enzymes work on specific foods. You need the right type of enzyme for the foods you want it to break down. Digestive enzymes are enzymes that break down food into usable material. The major different types of digestive enzymes are:

- amylase – breaks down carbohydrates, starches, and sugars which are prevalent in potatoes, fruits, vegetables, and many snack foods
- protease – breaks down proteins found in meats, nuts, eggs, and cheese
- lipase – breaks down fats found in most dairy products, nuts, oils, and meat
- cellulase – breaks down cellulose, plant fibre; not found in humans

Enzyme is a Biological Catalyst

Catalyst is a molecule that increases the rate of a chemical reaction without itself being permanently changed after the reaction.

Enzymes are proteins that catalyse (*i.e.*, increase or decrease the rates of) chemical reactions. In enzymatic reactions, the molecules at the beginning of the process are called substrates, and they are converted into different molecules, called the products. Almost all processes in a biological cell need enzymes to occur at significant rates. Since enzymes are selective for their substrates and speed up only a few reactions from among many possibilities, the set of enzymes made in a cell determines which metabolic pathways occur in that cell.

Like all catalysts, enzymes work by lowering the activation energy ($E_a^{\ddagger}$) for a reaction, thus dramatically increasing the rate of the reaction. As a result, products are formed faster and reactions reach their equilibrium state more rapidly. Most enzyme reaction rates are millions of times faster than those of comparable un-catalysed reactions. As with all catalysts, enzymes are not consumed by the reactions they catalyse, nor do they alter the equilibrium of these reactions. However, enzymes do differ from most other catalysts by being much more specific. Enzymes are known to catalyse about 4,000 biochemical reactions. A few RNA molecules called ribozymes also catalyse reactions, with an important example being some parts of the ribosome. Synthetic molecules called artificial enzymes also display enzyme-like catalysis.

Enzyme activity can be affected by other molecules. Inhibitors are molecules that decrease enzyme activity; activators are molecules that increase activity. Many drugs and poisons are enzyme inhibitors.

Activity is also affected by temperature, chemical environment (*e.g.*, pH), and the concentration of substrate. Some enzymes are used commercially, for example, in the synthesis of antibiotics. In addition, some household products use enzymes to speed up biochemical reactions (*e.g.*, enzymes in biological washing powders break down protein or fat stains on clothes; enzymes in meat tenderizers break down proteins, making the meat easier to chew).

Structures and Mechanisms

Enzymes are generally globular proteins and range from just 62 amino acid residues in size, for the monomer of 4-oxalocrotonate tautomerase, to over 2,500 residues in the animal fatty acid synthase. A small number of RNA-based biological catalysts exist, with the most common being the ribosome; these are referred to as either RNA-enzymes or ribozymes. The activities of enzymes are determined by their three-dimensional structure. However, although structure does determine function, predicting a novel enzyme's activity just from its structure is a very difficult problem that has not yet been solved.

Most enzymes are much larger than the substrates they act on, and only a small portion of the enzyme (around 3–4 amino acids) is directly involved in catalysis. The region that contains these catalytic residues, binds the substrate, and then carries out the reaction is known as the active site. Enzymes can also contain sites that bind cofactors, which are needed for catalysis. Some enzymes also have binding sites for small molecules, which are often direct or indirect products or substrates of the reaction catalysed. This binding can serve to increase or decrease the enzyme's activity, providing a means for feedback regulation.

Like all proteins, enzymes are long, linear chains of amino acids that fold to produce a three-dimensional product. Each unique amino acid sequence produces a specific structure, which has unique properties. Individual protein chains may sometimes group together to form a protein complex. Most enzymes can be denatured—that is, unfolded and inactivated—by heating or chemical denaturants, which disrupt the three-dimensional structure of the protein. Depending on the enzyme, denaturation may be reversible or irreversible.

Structures of enzymes in complex with substrates or substrate analogs during a reaction may be obtained using Time resolved crystallography methods.

Specificity

Enzymes are usually very specific as to which reactions they catalyse and the substrates that are involved in these reactions. Complementary shape, charge and hydrophilic/hydrophobic characteristics of enzymes and substrates are responsible for this specificity. Enzymes can also show impressive levels of stereospecificity, regioselectivity and chemoselectivity.

Some of the enzymes showing the highest specificity and accuracy are involved in the copying and expression of the genome. These enzymes have "proof-reading" mechanisms. Here, an enzyme such as DNA polymerase catalyzes a reaction in a first step and then checks that the product is correct in a second step. This two-step process results in average error rates of less than 1 error in 100 million reactions in high-fidelity mammalian polymerases. Similar proofreading mechanisms are also found in RNA polymerase, aminoacyl tRNA synthetases and ribosomes.

Some enzymes that produce secondary metabolites are described as promiscuous, as they can act on a relatively broad range of different substrates. It has been suggested that this broad substrate specificity is important for the evolution of new biosynthetic pathways.

"Lock and Key" Model

Enzymes are very specific, and it was suggested by the Nobel laureate organic chemist Emil Fischer in 1894 that this was because both the enzyme and the substrate possess specific complementary geometric shapes that fit exactly into one another. This is often referred to as "the lock and key" model. However, while this model explains enzyme specificity, it fails to explain the stabilization of the transition state that enzymes achieve. In 1958, Daniel Koshland suggested a modification to the lock and key model: since enzymes are rather flexible structures, the active site is continually reshaped by interactions with the substrate as the substrate interacts with the enzyme. As a result, the substrate does not simply bind to a rigid active site; the amino acid side chains which make up the active site are molded into the precise positions that enable the enzyme to perform its catalytic function. In some cases, such as glycosidases, the substrate molecule also changes shape slightly as it enters the active site. The active site continues to change until the substrate is completely bound, at which point the final shape and charge is determined. Induced fit may enhance the fidelity of molecular recognition in the presence of competition and noise via the conformational proofreading mechanism.

Mechanisms

Enzymes can act in several ways, all of which lower $\Delta G^{\ddagger}$:

- Lowering the activation energy by creating an environment in which the transition state is stabilized (e.g. straining the shape of a substrate—by binding the transition-state conformation of the substrate/product molecules, the enzyme distorts the bound substrate(s) into their transition state form, thereby reducing the amount of energy required to complete the transition).
- Lowering the energy of the transition state, but without distorting the substrate, by creating an environment with the opposite charge distribution to that of the transition state.
- Providing an alternative pathway. For example, temporarily reacting with the substrate to form an intermediate ES complex, which would be impossible in the absence of the enzyme.
- Reducing the reaction entropy change by bringing substrates together in the correct orientation to react. Considering $\Delta H^{\ddagger}$ alone overlooks this effect.
- Increases in temperatures speed up reactions. Thus, temperature increases help the enzyme function and develop the end product even faster. However, if heated too much, the enzyme's shape deteriorates and only when the temperature comes back to normal does the enzyme regain its shape. Some enzymes like thermolabile enzymes work best at low temperatures.

Interestingly, this entropic effect involves destabilization of the ground state, and its contribution to catalysis is relatively small.

Transition State Stabilization

The understanding of the origin of the reduction of $\Delta G^{\ddagger}$ requires one to find out how the enzymes can stabilize its transition state more than the transition state of the uncatalyzed reaction. Apparently, the most effective way for reaching large stabilization is the use of electrostatic effects, in particular, by having a relatively fixed polar environment that is oriented towards the charge distribution of the transition state. Such an environment does not exist in the uncatalyzed reaction in water.

Dynamics and Function

The internal dynamics of enzymes is linked to their mechanism of catalysis. Internal dynamics are the movement of parts of the enzyme's structure, such as individual amino acid residues, a group of amino acids, or even an entire protein domain. These movements occur at various time-scales ranging from femtoseconds to seconds. Networks of protein residues throughout an enzyme's structure can contribute to catalysis through dynamic motions. Protein motions are vital to many enzymes, but whether small and fast vibrations, or larger and slower conformational movements are more important depends on the type of reaction involved. However, although these movements are important in binding and releasing substrates and products, it is not clear if protein movements help to accelerate the chemical steps in enzymatic reactions. These new insights also have implications in understanding allosteric effects and developing new drugs.

Allosteric Modulation

Allosteric sites are sites on the enzyme that bind to molecules in the cellular environment. The sites form weak, noncovalent bonds with these molecules, causing a change in the conformation of the enzyme. This change in conformation translates to the active site, which then affects the reaction rate of the enzyme. Allosteric interactions can both inhibit and activate enzymes and are a common way that enzymes are controlled in the body.

Cofactors and Coenzymes

Cofactors

Some enzymes do not need any additional components to show full activity. However, others require non-protein molecules called cofactors to be bound for activity. Cofactors can be either inorganic (*e.g.*, metal ions and iron-sulphur clusters) or organic compounds (e.g., flavin and heme). Organic cofactors can be either prosthetic groups, which are tightly bound to an enzyme, or coenzymes, which are released from the enzyme's active site during the reaction. Coenzymes include NADH, NADPH and adenosine triphosphate. These molecules transfer chemical groups between enzymes.

An example of an enzyme that contains a cofactor is carbonic anhydrase, and is shown in the ribbon diagram above with a zinc cofactor bound as part of its active site. These tightly bound molecules are usually found in the active site and are involved in catalysis. For example, flavin and heme cofactors are often involved in redox reactions.

Enzymes that require a cofactor but do not have one bound are called *apoenzymes* or *apoproteins*. An apoenzyme together with its cofactor(s) is called a *holoenzyme* (this is the active form). Most cofactors are not covalently attached to an enzyme, but are very tightly bound. However, organic prosthetic groups can be covalently bound (*e.g.*, thiamine pyrophosphate in the enzyme pyruvate dehydrogenase). The term "holoenzyme" can also be applied to enzymes that contain multiple protein subunits, such as the DNA polymerases; here the holoenzyme is the complete complex containing all the subunits needed for activity.

Coenzymes

Coenzymes are small organic molecules that transport chemical groups from one enzyme to another. Some of these chemicals such as riboflavin, thiamine and folic acid are vitamins (compounds which cannot be synthesized by the body and must be acquired from the diet). The chemical groups carried include the hydride ion (H^-) carried by NAD or $NADP^+$, the phosphate group carried by adenosine triphosphate, the acetyl group carried by coenzyme A, formyl, methenyl or methyl groups carried by folic acid and the methyl group carried by S-adenosylmethionine.

Since coenzymes are chemically changed as a consequence of enzyme action, it is useful to consider coenzymes to be a special class of substrates, or second substrates, which are common to many different enzymes. For example, about 700 enzymes are known to use the coenzyme NADH.

Coenzymes are usually continuously regenerated and their concentrations maintained at a steady level inside the cell: for example, NADPH is regenerated through the pentose phosphate pathway and *S*-adenosylmethionine by methionine adenosyltransferase. This continuous regeneration means that even small amounts of coenzymes are used very intensively. For example, the human body turns over its own weight in ATP each day.

Thermodynamics

As all catalysts, enzymes do not alter the position of the chemical equilibrium of the reaction. Usually, in the presence of an enzyme, the reaction runs in the same direction as it would without the enzyme, just more quickly. However, in the absence of the enzyme, other possible uncatalyzed, "spontaneous" reactions might lead to different products, because in those conditions this different product is formed faster. Furthermore, enzymes can couple two or more reactions, so that a thermodynamically favourable reaction can be used to "drive" a thermodynamically unfavourable one. For example, the hydrolysis of ATP is often used to drive other chemical reactions.

Enzymes catalyse the forward and backward reactions equally. They do not alter the equilibrium itself, but only the speed at which it is reached. For example, carbonic anhydrase catalyzes its reaction in either direction depending on the concentration of its reactants.

$$CO_2 + H_2O \xrightarrow{\text{Carbonic anhydrase}} H_2CO_3$$

(in tissues; high CO_2 concentration)

$$H_2CO_3 \xrightarrow{\text{Carbonic anhydrase}} CO_2H_2O$$

(in lungs; low CO_2 concentration)

Nevertheless, if the equilibrium is greatly displaced in one direction, that is, in a very exergonic reaction, the reaction is *effectively* irreversible. Under these conditions the enzyme will, in fact, only catalyse the reaction in the thermodynamically allowed direction.

Kinetics

Enzyme kinetics is the investigation of how enzymes bind substrates and turn them into products. The rate data used in kinetic analyses are obtained from enzyme assays.

In 1902 Victor Henri proposed a quantitative theory of enzyme kinetics, but his experimental data were not useful because the significance of the hydrogen ion concentration was not yet appreciated. After Peter Lauritz Sørensen had defined the logarithmic pH-scale and introduced the concept of buffering in 1909 the German chemist Leonor Michaelis and his Canadian postdoc Maud Leonora Menten repeated Henri's experiments and confirmed his equation which is referred to as Henri-Michaelis-Menten kinetics (sometimes also Michaelis-Menten kinetics). Their work was further developed by G. E. Briggs and J. B. S. Haldane, who derived kinetic equations that are still widely used today. The major contribution of Henri was to think of enzyme reactions in two stages. In the first, the substrate binds reversibly to the enzyme, forming the enzyme-substrate complex. This is sometimes called the Michaelis complex. The enzyme then catalyzes the chemical step in the reaction and releases the product.

Enzymes can catalyse up to several million reactions per second. For example, the uncatalyzed decarboxylation of orotidine 5'-monophosphate has a half life of 78 million years. However, when the enzyme orotidine 5'-phosphate decarboxylase is added, the same process takes just 25 milliseconds. Enzyme rates depend on solution conditions and substrate concentration. Conditions that denature the protein abolish enzyme activity, such as high temperatures, extremes of pH or high salt concentrations, while raising substrate concentration tends to increase activity. To find the maximum speed of an enzymatic reaction, the substrate concentration is increased until a constant rate of product formation is seen. This is shown in the saturation curve on the right. Saturation happens because, as substrate concentration increases, more and more of the free enzyme is converted into the substrate-bound ES form. At the maximum velocity (V_{max}) of the enzyme, all the enzyme active sites are bound to substrate, and the amount of ES complex is the same as the total amount of enzyme. However, V_{max} is only one kinetic constant of enzymes. The amount of substrate needed to achieve a given rate of reaction is also important. This is given by the Michaelis-Menten constant (K_m), which is the substrate concentration required for an enzyme to reach one-half its maximum velocity. Each enzyme has a characteristic K_m for a given substrate, and this can show how tight the binding of the substrate is to the enzyme. Another useful constant is k_{cat}, which is the number of substrate molecules handled by one active site per second.

The efficiency of an enzyme can be expressed in terms of k_{cat}/K_m. This is also called the specificity constant and incorporates the rate constants for all steps in the reaction. Because the specificity constant reflects both affinity and catalytic ability, it is useful for comparing different enzymes against each other, or the same enzyme with different substrates. The theoretical maximum for the specificity constant is called the diffusion limit and is about 10^8 to 10^9 (M^{-1} s^{-1}). At this point every collision of the enzyme with its substrate will result in catalysis, and the rate of product formation is not limited by the reaction rate but by the diffusion rate. Enzymes with this property are called *catalytically perfect* or *kinetically perfect*. Example of such enzymes are triose-phosphate isomerase, carbonic anhydrase, acetylcholinesterase, catalase, fumarase, β-lactamase, and superoxide dismutase.

Michaelis-Menten kinetics relies on the law of mass action, which is derived from the assumptions of free diffusion and thermodynamically driven random collision. However, many biochemical or cellular processes deviate significantly from these conditions, because of macromolecular crowding, phase-separation of the enzyme/substrate/ product, or one or two-dimensional molecular movement. In these situations, a fractal Michaelis-Menten kinetics may be applied.

Some enzymes operate with kinetics which are faster than diffusion rates, which would seem to be impossible. Several mechanisms have been invoked to explain this phenomenon. Some proteins are believed to accelerate catalysis by drawing their substrate in and pre-orienting them by using dipolar electric fields. Other models invoke a quantum-mechanical tunnelling explanation, whereby a proton or an electron can tunnel through activation barriers, although for proton tunnelling this model remains somewhat controversial. Quantum tunnelling for protons has been observed in tryptamine. This suggests that enzyme catalysis may be more accurately characterized as "through the barrier" rather than the traditional model, which requires substrates to go "over" a lowered energy barrier.

Inhibition

Enzyme inhibitors are molecules that bind to enzymes and decrease their activity. Since blocking an enzyme's activity can kill a pathogen or correct a metabolic imbalance, many drugs are enzyme inhibitors. They are also used as herbicides and pesticides. Not all molecules that bind to enzymes are inhibitors; *enzyme activators* bind to enzymes and increase their enzymatic activity.

The binding of an inhibitor can stop a substrate from entering the enzyme's active site and/or hinder the enzyme from catalysing its reaction. Inhibitor binding is either reversible or irreversible. Irreversible inhibitors usually react with the enzyme and change it chemically. These inhibitors modify key amino acid residues needed for enzymatic activity. In contrast, reversible inhibitors bind non-covalently and different types of inhibition are produced depending on whether these inhibitors bind the enzyme, the enzyme-substrate complex, or both.

Many drug molecules are enzyme inhibitors, so their discovery and improvement is an active area of research in biochemistry and pharmacology. A medicinal enzyme inhibitor is often judged by its specificity (its lack of binding to other proteins) and its potency (its dissociation constant, which indicates the concentration needed to inhibit the enzyme). A high specificity and potency ensure that a drug will have few side effects and thus low toxicity. Enzyme inhibitors also occur naturally and are involved in the regulation of metabolism. For example, enzymes in a metabolic pathway can be inhibited by downstream products. This type of negative feedback slows flux through a pathway when the products begin to build up and is an important way to maintain homeostasis in a cell. Other cellular enzyme inhibitors are proteins that specifically bind to and inhibit an enzyme target. This can help control enzymes that may be damaging to a cell, such as proteases or nucleases; a well-characterised example is the ribonuclease inhibitor, which binds to ribonucleases in one of the tightest known protein–protein interactions. Natural enzyme inhibitors can also be poisons and are used as defences against predators or as ways of killing prey.

Reversible Inhibitors

Types of Reversible Inhibitors

Reversible inhibitors bind to enzymes with non-covalent interactions such as hydrogen bonds, hydrophobic interactions and ionic bonds. Multiple weak bonds between the inhibitor and the active site combine to produce strong and specific binding. In contrast to substrates and irreversible inhibitors, reversible inhibitors generally do not undergo chemical reactions when bound to the enzyme and can be easily removed by dilution or dialysis.

There are four kinds of reversible enzyme inhibitors. They are classified according to the effect of varying the concentration of the enzyme's substrate on the inhibitor.

- In competitive inhibition, the substrate and inhibitor cannot bind to the enzyme at the same time, as shown in the figure on the left. This usually results from the inhibitor having an affinity for the active site of an enzyme where the substrate also binds; the substrate and inhibitor *compete* for access to the enzyme's active site. This type of inhibition can be overcome by sufficiently high concentrations of substrate, i.e., by out-competing the inhibitor. Competitive inhibitors are often similar in structure to the real substrate.
- In uncompetitive inhibition, the inhibitor binds only to the substrate-enzyme complex, it should not be confused with non-competitive inhibitors. Both Vmax and Km decrease (maximum velocity decreases as a result of removing activated complex

while binding efficiency increases as a result of Le Chatelier's principle).

- In mixed inhibition, the inhibitor can bind to the enzyme at the same time as the enzyme's substrate. However, the binding of the inhibitor affects the binding of the substrate, and vice versa. This type of inhibition can be reduced, but not overcome by increasing concentrations of substrate. Although it is possible for mixed-type inhibitors to bind in the active site, this type of inhibition generally results from an allosteric effect where the inhibitor binds to a different site on an enzyme. Inhibitor binding to this allosteric site changes the conformation (i.e., tertiary structure or three-dimensional shape) of the enzyme so that the affinity of the substrate for the active site is reduced.
- Non-competitive inhibition is a form of mixed inhibition where the binding of the inhibitor to the enzyme reduces its activity but does not affect the binding of substrate. As a result, the extent of inhibition depends only on the concentration of the inhibitor.

Quantitative Description of Reversible Inhibition

Reversible inhibition can be described quantitatively in terms of the inhibitor's binding to the enzyme and to the enzyme–substrate complex, and its effects on the kinetic constants of the enzyme. In the classic Michaelis–Menten scheme below, an enzyme (E) binds to its substrate (S) to form the enzyme–substrate complex ES. Upon catalysis, this complex breaks down to release product P and free enzyme. The inhibitor (I) can bind to either E or ES with the dissociation constants K_i or K_i', respectively.

When an enzyme has multiple substrates, inhibitors can show different types of inhibition depending on which substrate is considered. This results from the active site containing two different binding sites within the active site, one for each substrate. For example, an inhibitor might compete with substrate A for the first binding site, but be a non-competitive inhibitor with respect to substrate B in the second binding site.

Measuring the Dissociation Constants of a Reversible Inhibitor

As noted above, an enzyme inhibitor is characterised by its two dissociation constants, K_i and K_i', to the enzyme and to the enzyme-substrate complex, respectively. The enzyme-inhibitor constant K_i can be measured directly by various methods; one extremely accurate method is isothermal titration calorimetry, in which the inhibitor is titrated into a solution of enzyme and the heat released or absorbed is measured. However, the other dissociation constant K_i' is difficult to measure directly, since the enzyme-substrate complex is short-lived and undergoing a chemical reaction to form the product. Hence, K_i' is usually measured indirectly, by observing the enzyme activity under various substrate and inhibitor concentrations, and fitting the data to a modified Michaelis–Menten equation

$$V = \frac{V_{\max}[S]}{\alpha K_m + \alpha'[S]} = \frac{(1/\alpha')V_{\max}[S]}{(\alpha/\alpha')K_m + [S]}$$

where the modifying factors α and α' are defined by the inhibitor concentration and its two dissociation constants

$$\alpha = 1 + \frac{[1]}{K_i}$$

$$\alpha' = 1 + \frac{[I]}{K_i'}.$$

Thus, in the presence of the inhibitor, the enzyme's effective K_m and V_{max} become $(\alpha/\alpha')K_m$ and $(1/\alpha')V_{max}$, respectively. However, the modified Michaelis-Menten equation assumes that binding of the inhibitor to the enzyme has reached equilibrium, which may be a very slow process for inhibitors with sub-nanomolar dissociation constants.

In these cases, it is usually more practical to treat the tight-binding inhibitor as an irreversible inhibitor; however, it can still be possible to estimate K_i' kinetically if K_i is measured independently.

The effects of different types of reversible enzyme inhibitors on enzymatic activity can be visualized using graphical representations of the Michaelis–Menten equation, such as Lineweaver–Burk and Eadie-Hofstee plots. For example, in the Lineweaver–Burk plots at the right, the competitive inhibition lines intersect on the *y*-axis, illustrating that such inhibitors do not affect V_{max}. Similarly, the non-competitive inhibition lines intersect on the *x*-axis, showing these inhibitors do not affect K_m. However, it can be difficult to estimate K_i and K_i' accurately from such plots, so it is advisable to estimate these constants using more reliable nonlinear regression methods, as described above.

Special Cases

- The mechanism of partially competitive inhibition is similar to that of non-competitive, except that the EIS complex has catalytic activity, which may be lower or even higher (partially competitive activation) than that of the enzyme–substrate (ES) complex. This inhibition typically displays a lower V_{max}, but an unaffected K_m value.
- Uncompetitive inhibition occurs when the inhibitor binds only to the enzyme–substrate complex, not to the free enzyme; the EIS complex is catalytically inactive. This mode of inhibition is rare and causes a decrease in both V_{max} and the K_m value.
- Substrate and product inhibition is where either the substrate or product of an enzyme reaction inhibit the enzyme's activity. This inhibition may follow the competitive, uncompetitive or mixed patterns. In substrate inhibition there is a progressive decrease in activity at high substrate concentrations. This may indicate the existence of two substrate-binding sites in the enzyme. At low substrate, the high-affinity site is occupied and normal kinetics are followed. However, at higher concentrations, the second inhibitory site becomes occupied, inhibiting the enzyme. Product inhibition is often a regulatory feature in metabolism and can be a form of negative feedback.
- Slow-tight inhibition occurs when the initial enzyme–inhibitor complex EI undergoes isomerisation to a second more tightly held complex, EI*, but the overall inhibition process is reversible. This manifests itself as slowly increasing enzyme inhibition. Under these conditions, traditional Michaelis–Menten kinetics give a false value for K_i, which is time–dependent. The true value of K_i can be obtained through more complex analysis of the on (k_{on}) and off (k_{off}) rate constants for inhibitor association.

Examples of Reversible Inhibitors

As enzymes have evolved to bind their substrates tightly, and most reversible inhibitors bind in the active site of enzymes, it is unsurprising that some of these inhibitors are strikingly similar in structure to the substrates of their targets. An example of these substrate mimics are the protease inhibitors, a very successful class of antiretroviral drugs used to treat HIV. The structure of ritonavir, a protease inhibitor based on a peptide and containing three peptide bonds, is shown on the right. As this drug resembles the protein that is the substrate of the HIV protease, it competes with this substrate in the enzyme's active site.

Enzyme inhibitors are often designed to mimic the transition state or intermediate of an enzyme-catalysed reaction. This ensures that the inhibitor exploits the transition state stabilising effect of the enzyme, resulting in a better binding affinity (lower K_i) than substrate-based designs. An example of such a transition state inhibitor is the antiviral drug oseltamivir; this drug mimics the planar nature of the ring oxonium ion in the reaction of the viral enzyme neuraminidase.

However, not all inhibitors are based on the structures of substrates. For example, the structure of another HIV protease inhibitor tipranavir is shown on the left. This molecule is not based on a peptide and has no obvious structural similarity to a protein substrate. These non-peptide inhibitors can be more stable than inhibitors containing peptide bonds, because they will not be substrates for peptidases and are less likely to be degraded.

In drug design it is important to consider the concentrations of substrates to which the target enzymes are exposed. For example, some protein kinase inhibitors have chemical structures that are similar to adenosine triphosphate, one of the substrates of these enzymes. However, drugs that are simple competitive inhibitors will have to compete with the high concentrations of ATP in the cell. Protein kinases can also be inhibited by competition at the binding sites where the kinases interact with their substrate proteins, and most proteins are present inside cells at concentrations much lower than the concentration of ATP. As a consequence, if two protein kinase inhibitors both bind in the active site with similar affinity, but only one has to compete with ATP, then the competitive inhibitor at the protein-binding site will inhibit the enzyme more effectively.

Irreversible Inhibitors

Types of Irreversible Inhibition

Irreversible inhibitors usually covalently modify an enzyme, and inhibition cannot therefore be reversed. Irreversible inhibitors often contain reactive functional groups such as nitrogen mustards, aldehydes, haloalkanes, alkenes, Michael acceptors, phenyl sulphonates, or fluorophosphonates. These electrophilic groups react with amino acid side chains to form covalent adducts. The residues modified are those with side chains containing nucleophiles such as hydroxyl or sulphhydryl groups; these include the amino acids serine (as in DFP, right), cysteine, threonine or tyrosine.

Irreversible inhibition is different from irreversible enzyme inactivation. Irreversible inhibitors are generally specific for one class of enzyme and do not inactivate all proteins; they do not function by destroying protein structure but by specifically altering the active site of their target. For example, extremes of pH or temperature usually cause denaturation of all protein structure, but this is a non-specific effect. Similarly, some non-specific chemical treatments destroy protein structure: for example, heating in concentrated hydrochloric acid will hydrolyse the peptide bonds holding proteins together, releasing free amino acids.

Irreversible inhibitors display time-dependent inhibition and their potency therefore cannot be characterised by an IC_{50} value. This is because the amount of active enzyme at a given concentration of irreversible inhibitor will be different depending on how long the inhibitor is pre-incubated with the enzyme. Instead, $k_{obs}/[I]$ values are used, wherek_{obs} is the observed pseudo-first order rate of inactivation (obtained by plotting the log of % activity vs. time) and $[I]$ is the concentration of inhibitor. The $k_{obs}/[I]$ parameter is valid as long as the inhibitor does not saturate binding with the enzyme (in which case $k_{obs} = k_{inact}$).

Analysis of Irreversible Inhibition

The irreversible inhibitors form a reversible non-covalent complex with the enzyme (EI or ESI) and this then reacts to produce the covalently modified "dead-end complex" EI*. The rate at which EI* is formed is called the inactivation rate or k_{inact}. Since formation of EI may compete with ES, binding of irreversible inhibitors can be prevented by competition either with substrate or with a second, reversible inhibitor. This protection effect is good evidence of a specific reaction of the irreversible inhibitor with the active site.

The binding and inactivation steps of this reaction are investigated by incubating the enzyme with inhibitor and assaying the amount of activity remaining over time. The activity will be decrease in a time-dependent manner, usually following exponential decay. Fitting these data to a rate equation gives the rate of inactivation at this concentration of inhibitor. This is done at several different concentrations of inhibitor. If a reversible EI complex is involved the inactivation rate will be saturable and fitting this curve will give k_{inact} and K_i. Another method that is widely used in these analyses is mass spectrometry. Here, accurate measurement of the mass of the unmodified native enzyme and the inactivated enzyme gives the increase in mass caused by reaction with the inhibitor and shows the stoichiometry of the reaction. This is usually done using a MALDI-TOF mass spectrometer. In a complementary technique, peptide mass fingerprinting involves digestion of the native and modified protein with a protease such as trypsin. This will produce a set of peptides that can be analysed using a mass spectrometer. The peptide that changes in mass after reaction with the inhibitor will be the one that contains the site of modification.

Special Cases

Not all irreversible inhibitors form covalent adducts with their enzyme targets. Some reversible inhibitors bind so tightly to their target enzyme that they are essentially irreversible. These tight-binding inhibitors may show kinetics similar to covalent irreversible inhibitors. In these cases, some of these inhibitors rapidly bind to the enzyme in a low-affinity EI complex and this then undergoes a slower rearrangement to a very tightly bound EI* complex. This kinetic behaviour is called slow-binding. This slow rearrangement after binding often involves a conformational change as the enzyme "clamps down" around the inhibitor molecule. Examples of slow-binding inhibitors include some important drugs, such methotrexate, allopurinol, and the activated form of acyclovir.

Examples of Irreversible Inhibitors

Diisopropylfluorophosphate (DFP) is shown as an example of an irreversible protease inhibitor in the figure above right. The enzyme hydrolyses the phosphorus–fluorine bond, but the phosphate residue remains bound to the serine in the active site, deactivating it. Similarly, DFP also reacts with the active site of acetylcholine esterase in the synapses of neurons, and consequently is a potent neurotoxin, with a lethal dose of less than 100 mg.

Suicide inhibition is an unusual type of irreversible inhibition where the enzyme converts the inhibitor into a reactive form in its active site. An example is the inhibitor of polyamine biosynthesis, α-difluoromethylornithine or DFMO, which is an analogue of the amino acid ornithine, and is used to treat African trypanosomiasis (sleeping sickness). Ornithine decarboxylase can catalyse the decarboxylation of DFMO instead of ornithine, as shown above. However, this decarboxylation reaction is followed by the elimination of a fluorine atom, which converts this catalytic intermediate into a conjugated imine, a highly electrophilic species. This reactive form of DFMO then reacts with either a cysteine or lysine residue in the active site to irreversibly inactivate the enzyme.

Since irreversible inhibition often involves the initial formation of a non-covalent EI complex, it is sometimes possible for an inhibitor to bind to an enzyme in more than one way. For example, in the figure showing trypanothione reductase from the human protozoan parasite *Trypanosoma cruzi*, two molecules of an inhibitor called *quinacrine mustard* are bound in its active site. The top molecule is bound reversibly, but the lower one is bound covalently as it has reacted with an amino acid residue through its nitrogen mustard group.

Discovery and Design of Inhibitors

New drugs are the products of a long drug development process, the first step of which is often the discovery of a new enzyme inhibitor. In the past the only way to discover these new inhibitors was by trial and error: screening huge libraries of compounds against a target enzyme and hoping that some useful leads would emerge. This brute force approach is still successful and has even been extended by combinatorial chemistry approaches that quickly produce large numbers of novel compounds and high-throughput screening technology to rapidly screen these huge chemical libraries for useful inhibitors.

More recently, an alternative approach has been applied: rational drug design uses the three-dimensional structure of an enzyme's active site to predict which molecules might be inhibitors. These predictions are then tested and one of these tested compounds may be a novel inhibitor. This new inhibitor is then used to try to obtain a structure of the enzyme in an inhibitor/enzyme complex to show how the molecule is binding to the active site, allowing changes to be made to the inhibitor to try to optimise binding. This test and improve cycle is then repeated until a sufficiently potent inhibitor is produced. Computer-based methods of predicting the affinity of an inhibitor for an enzyme are also being developed, such as molecular docking and molecular mechanics.

Uses of Inhibitors

Enzyme inhibitors are found in nature and are also designed and produced as part of pharmacology and biochemistry. Natural poisons are often enzyme inhibitors that have evolved to defend a plant or animal against predators. These natural toxins include some of the most poisonous compounds known. Artificial inhibitors are often used as drugs, but can also be insecticides such as malathion, herbicides such as glyphosate, or disinfectants such as triclosan.

Chemotherapy

The most common uses for enzyme inhibitors are as drugs to treat disease. Many of these inhibitors target a human enzyme and aim to correct a pathological condition. However, not all drugs are enzyme inhibitors. Some, such as anti-epileptic drugs, alter enzyme activity by causing more or less of the enzyme to be produced. These effects are called enzyme induction and inhibition and are alterations in gene expression, which is unrelated to the type of enzyme inhibition. Other drugs interact with cellular targets that are not enzymes, such as ion channels or membrane receptors.

An example of a medicinal enzyme inhibitor is sildenafil (Viagra), a common treatment for male erectile dysfunction. This compound is a potent inhibitor of cGMP specific phosphodiesterase type 5, the enzyme that degrades the signalling molecule cyclic guanosine monophosphate. This signalling molecule triggers smooth muscle relaxation and allows blood flow into the corpus cavernosum, which causes an erection. Since the drug decreases the activity of the enzyme that halts the signal, it makes this signal last for a longer period of time.

Another example of the structural similarity of some inhibitors to the substrates of the enzymes they target is seen in the figure comparing the drug methotrexate to folic acid. Folic acid is a substrate of dihydrofolate reductase, an enzyme involved in making nucleotides that is potently inhibited by methotrexate. Methotrexate blocks the action of dihydrofolate reductase and thereby halts the production of nucleotides. This block of nucleotide biosynthesis is more toxic to rapidly growing cells than non-dividing cells, since a rapidly-growing cell has to carry out DNA replication, therefore methotrexate is often used in cancer chemotherapy.

Drugs also are used to inhibit enzymes needed for the survival of pathogens. For example, bacteria are surrounded by a thick cell wall made of a net-like polymer called peptidoglycan. Many antibiotics such as penicillin and vancomycin inhibit the enzymes that produce and then cross-link the strands of this polymer together. This causes the cell wall to lose strength and the bacteria to burst. In the figure, a molecule of penicillin (shown in a ball-and-stick form) is shown bound to its target, the transpeptidase from the bacteria *Streptomyces* R61 (the protein is shown as a ribbon-diagram).

Drug design is facilitated when an enzyme that is essential to the pathogen's survival is absent or very different in humans. In the example above, humans do not make peptidoglycan, therefore inhibitors of this process are selectively toxic to bacteria. Selective toxicity is also produced in antibiotics by exploiting differences in the structure of the ribosomes in bacteria, or how they make fatty acids.

Metabolic Control

Enzyme inhibitors are also important in metabolic control. Many metabolic pathways in the cell are inhibited by metabolites that control enzyme activity through allosteric regulation or substrate inhibition. A good example is the allosteric regulation of the glycolytic pathway. This catabolic pathway consumes glucose and produces ATP, NADH and pyruvate. A key step for the regulation of glycolysis is an early reaction in the pathway catalysed by phosphofructokinase-1 (PFK1). When ATP levels rise, ATP binds an allosteric site in PFK1 to decrease the rate of the enzyme reaction; glycolysis is inhibited and ATP production falls. This negative feedback control helps maintain a steady concentration of ATP in the cell. However, metabolic pathways are not just regulated through inhibition since enzyme activation is equally important. With respect to PFK1, fructose 2,6-bisphosphate and ADP are examples of metabolites that are allosteric activators.

Physiological enzyme inhibition can also be produced by specific protein inhibitors. This mechanism occurs in the pancreas, which synthesises many digestive precursor enzymes known as zymogens. Many of these are activated by the trypsin protease, so it is important to inhibit the activity of trypsin in the pancreas to prevent the organ from digesting itself. One way in which the activity of trypsin is controlled is the production of a specific and potent trypsin inhibitor protein in the pancreas. This inhibitor binds tightly to trypsin, preventing the trypsin activity that would otherwise be detrimental to the organ. Although the trypsin inhibitor is a protein, it avoids being hydrolysed as a substrate by the protease by excluding water from trypsin's active site and destabilising the transition state. Other examples of physiological enzyme inhibitor proteins include the barstar inhibitor of the bacterial ribonuclease barnase and the inhibitors of protein phosphatases.

Pesticides and Herbicides

Many herbicides and pesticides are enzyme inhibitors. Acetylcholinesterase (AChE) is an enzyme found in animals from insects to humans. It is essential to nerve cell function through its mechanism of breaking down the neurotransmitter acetylcholine into its constituents, acetate and choline. This is somewhat unique among neurotransmitters as most, including serotonin, dopamine, and norepinephrine, are absorbed from the synaptic cleft rather than cleaved. A large number of AChE inhibitors are used in both medicine and agriculture. Reversible competitive inhibitors, such as edrophonium, physostigmine, and neostigmine, are used in the treatment of myasthenia gravis and in anaesthesia. The carbamate pesticides are also examples of reversible AChE inhibitors. The organophosphate insecticides such as malathion, parathion, and chlorpyrifos irreversibly inhibit acetylcholinesterase.

The herbicide glyphosate is an inhibitor of 3-phosphoshikimate 1-carboxyvinyltransferase, other herbicides, such as the sulphonylureas inhibit the enzyme acetolactate synthase. Both these enzymes are needed for plants to make branched-chain amino acids. Many other enzymess are inhibited by herbicides, including enzymes needed for the biosynthesis of lipids and carotenoids and the processes of photosynthesis and oxidative phosphorylation.

Natural Poisons

Animals and plants have evolved to synthesise a vast array of poisonous products including secondary metabolites, peptides and proteins that can act as inhibitors. Natural toxins are usually small organic molecules and are so diverse that there are probably natural inhibitors for most metabolic processes. The metabolic processes targeted by natural poisons encompass more than enzymes in metabolic pathways and can also include the inhibition of receptor, channel and structural protein functions in a cell. For example, paclitaxel (taxol), an organic molecule found in the Pacific yew tree, binds tightly to tubulin dimers and inhibits their assembly into microtubules in the cytoskeleton.

Many natural poisons act as neurotoxins that can cause paralysis leading to death and have functions for defence against predators or in hunting and capturing prey. Some of these natural inhibitors, despite their toxic attributes, are valuable for therapeutic uses at lower doses. An example of a neurotoxin are the glycoalkaloids, from the plant species in the *Solanaceae* family (includes potato, tomato and eggplant), that are acetylcholinesterase inhibitors. Inhibition of this enzyme causes an uncontrolled increase in the acetylcholine neurotransmitter, muscular paralysis and then death. Neurotoxicity can also result from the inhibition of receptors; for example, atropine from deadly nightshade (*Atropa belladonna*) that functions as a competitive antagonist of the muscarinic acetylcholine receptors.

Although many natural toxins are secondary metabolites, these poisons also include peptides and proteins. An example of a toxic peptide is alpha-amanitin, which is found in relatives of the death cap mushroom. This is a potent enzyme inhibitor, in this case preventing the RNA polymerase II enzyme from transcribing DNA. The algal toxin microcystin is also a peptide and is an inhibitor of protein phosphatases. This toxin can contaminate water supplies after algal blooms and is a known carcinogen that can also cause acute liver hemorrhage and death at higher doses. Proteins can also be natural poisons or antinutrients, such as the trypsin inhibitors that are found in some legumes. A less common class of toxins are toxic enzymes: these act as irreversible inhibitors of their target enzymes and work by chemically modifying their substrate enzymes. An example is ricin, an extremely potent protein toxin found in castor oil beans. This enzyme is a glycosidase that inactivates ribosomes. Since ricin is a catalytic irreversible inhibitor, this allows just a single molecule of ricin to kill a cell.

Enzyme reaction rates can be decreased by various types of enzyme inhibitors.

Competitive Inhibition

In competitive inhibition, the inhibitor and substrate compete for the enzyme (i.e., they can not bind at the same time). Often competitive inhibitors strongly resemble the real substrate of the enzyme. For example, methotrexate is a competitive inhibitor of the enzyme dihydrofolate reductase, which catalyzes the reduction of dihydrofolate to tetrahydrofolate.

The similarity between the structures of folic acid and this drug. Note that binding of the inhibitor need *not* be to the substrate binding site (as frequently stated), if binding of the inhibitor changes the conformation of the enzyme to prevent substrate binding and *vice versa*. In competitive inhibition the maximal velocity of the reaction is not changed, but higher substrate concentrations are required to reach a given velocity, increasing the apparent K_m.

Uncompetitive Inhibition

In uncompetitive inhibition the inhibitor can not bind to the free enzyme, but only to the ES-complex. The EIS-complex thus formed is enzymatically inactive. This type of inhibition is rare, but may occur in multimeric enzymes.

Non-competitive Inhibition

Non-competitive inhibitors can bind to the enzyme at the same time as the substrate, i.e. they *never* bind to the active site. Both the EI and EIS complexes are enzymatically inactive. Because the inhibitor can not be driven from the enzyme by higher substrate concentration (in contrast to competitive inhibition), the apparent V_{max} changes. But because the substrate can still bind to the enzyme, the K_m stays the same.

Mixed Inhibition

This type of inhibition resembles the non-competitive, except that the EIS-complex has residual enzymatic activity.

In many organisms inhibitors may act as part of a feedback mechanism. If an enzyme produces too much of one substance in the organism, that substance may act as an inhibitor for the enzyme at the beginning of the pathway that produces it, causing production of the substance to slow down or stop when there is sufficient amount. This is a form of negative feedback. Enzymes which are subject to this form of regulation are often multimeric and have allosteric binding sites for regulatory substances. Their substrate/velocity plots are not hyperbolar, but sigmoidal (S-shaped).

Irreversible inhibitors react with the enzyme and form a covalent adduct with the protein. The inactivation is irreversible. These compounds include eflornithine a drug used to treat the parasitic disease sleeping sickness. Penicillin and Aspirin also act in this manner. With these drugs, the compound is bound in the active site and the enzyme then converts the inhibitor into an activated form that reacts irreversibly with one or more amino acid residues.

Uses of Inhibitors

Since inhibitors modulate the function of enzymes they are often used as drugs. An common example of an inhibitor that is used as a drug is aspirin, which inhibits the COX-1 and COX-2 enzymes that produce the inflammation messenger prostaglandin, thus suppressing pain and inflammation. However, other enzyme inhibitors are poisons. For example, the poison cyanide is an irreversible enzyme inhibitor that combines with the copper and iron in the active site of the enzyme cytochrome c oxidase and blocks cellular respiration.

Biological Function

Enzymes serve a wide variety of functions inside living organisms. They are indispensable for signal transduction and cell regulation, often via kinases and phosphatases. They also generate movement, with myosin hydrolysing ATP to generate muscle contraction and also moving cargo around the cell as part of the cytoskeleton.

Other ATPases in the cell membrane are ion pumps involved in active transport. Enzymes are also involved in more exotic functions, such as luciferase generating light in fireflies. Viruses can also contain enzymes for infecting cells, such as the HIV integrase and reverse transcriptase, or for viral release from cells, like the influenza virus neuraminidase.

An important function of enzymes is in the digestive systems of animals. Enzymes such as amylases and proteases break down large molecules (starch or proteins, respectively) into smaller ones, so they can be absorbed by the intestines. Starch molecules, for example, are too large to be absorbed from the intestine, but enzymes hydrolyse the starch chains into smaller molecules such as maltose and eventually glucose, which can then be absorbed.

Different enzymes digest different food substances. In ruminants which have herbivorous diets, microorganisms in the gut produce another enzyme, cellulase to break down the cellulose cell walls of plant fibre.

Several enzymes can work together in a specific order, creating metabolic pathways. In a metabolic pathway, one enzyme takes the product of another enzyme as a substrate. After the catalytic reaction, the product is then passed on to another enzyme. Sometimes more than one enzyme can catalyse the same reaction in parallel, this can allow more complex regulation: with for example a low constant activity being provided by one enzyme but an inducible high activity from a second enzyme.

Enzymes determine what steps occur in these pathways. Without enzymes, metabolism would neither progress through the same steps, nor be fast enough to serve the needs of the cell. Indeed, a metabolic pathway such as glycolysis could not exist independently of enzymes. Glucose, for example, can react directly with ATP to become phosphorylated at one or more of its carbons.

In the absence of enzymes, this occurs so slowly as to be insignificant. However, if hexokinase is added, these slow reactions continue to take place except that phosphorylation at carbon 6 occurs so rapidly that if the mixture is tested a short time later, glucose-6-phosphate is found to be the only significant product. Consequently, the network of metabolic pathways within each cell depends on the set of functional enzymes that are present.

Control of Activity

There are five main ways that enzyme activity is controlled in the cell.

1. Enzyme production (transcription and translation of enzyme genes) can be enhanced or diminished by a cell in response to changes in the cell's environment. This form of gene regulation is called enzyme induction and inhibition. For example, bacteria may become resistant to antibiotics such as penicillin because enzymes called beta-lactamases are induced that hydrolyse the crucial beta-lactam ring within the penicillin molecule. Another example are enzymes in the liver called cytochrome P450 oxidases, which are important in drug metabolism. Induction or inhibition of these enzymes can cause drug interactions.
2. Enzymes can be compartmentalized, with different metabolic pathways occurring in different cellular compartments. For example, fatty acids are synthesized by one set of enzymes in the cytosol, endoplasmic reticulum and the Golgi apparatus and used by a different set of enzymes as a source of energy in the mitochondrion, through b-oxidation.
3. Enzymes can be regulated by inhibitors and activators. For example, the end product(s) of a metabolic pathway are often inhibitors for one of the first enzymes of the pathway (usually the first irreversible step, called *committed step*), thus regulating the amount of end product made by the pathways. Such a regulatory mechanism is called a negative feedback mechanism, because the amount of the end product produced is regulated

by its own concentration. Negative feedback mechanism can effectively adjust the rate of synthesis of intermediate metabolites according to the demands of the cells. This helps allocate materials and energy economically, and prevents the manufacture of excess end products. The control of enzymatic action helps to maintain a stable internal environment in living organisms.

4. Enzymes can be regulated through post-translational modification. This can include phosphorylation, myristoylation and glycosylation. For example, in the response to insulin, the phosphorylation of multiple enzymes, including glycogen synthase, helps control the synthesis or degradation of glycogen and allows the cell to respond to changes in blood sugar. Another example of post-translational modification is the cleavage of the polypeptide chain. Chymotrypsin, a digestive protease, is produced in inactive form as chymotrypsinogen in the pancreas and transported in this form to the stomach where it is activated. This stops the enzyme from digesting the pancreas or other tissues before it enters the gut. This type of inactive precursor to an enzyme is known as a zymogen.
5. Some enzymes may become activated when localized to a different environment (e.g. from a reducing (cytoplasm) to an oxidizing (periplasm) environment, high pH to low pH etc.). For example, hemagglutinin in the influenza virus is activated by a conformational change caused by the acidic conditions, these occur when it is taken up inside its host cell and enters the lysosome.

Involvement in Disease

Since the tight control of enzyme activity is essential for homeostasis, any malfunction (mutation, overproduction, underproduction or deletion) of a single critical enzyme can lead to a genetic disease. The importance of enzymes is shown by the fact that a lethal illness can be caused by the malfunction of just one type of enzyme out of the thousands of types present in our bodies.

One example is the most common type of phenylketonuria. A mutation of a single amino acid in the enzyme phenylalanine hydroxylase, which catalyzes the first step in the degradation of phenylalanine, results in build-up of phenylalanine and related products. This can lead to mental retardation if the disease is untreated. Another example is when germline mutations in genes coding for DNA repair enzymes cause hereditary cancer syndromes such as xeroderma pigmentosum. Defects in these enzymes cause cancer since the body is less able to repair mutations in the genome. This causes a slow accumulation of mutations and results in the development of many types of cancer in the sufferer.

Naming Conventions

An enzyme's name is often derived from its substrate or the chemical reaction it catalyzes, with the word ending in *-ase*. Examples are lactase, alcohol dehydrogenase and DNA polymerase. This may result in different enzymes, called isozymes, with the same function having the same basic name.

Isoenzymes have a different amino acid sequence and might be distinguished by their optimal pH, kinetic properties or immunologically. Furthermore, the normal physiological reaction an enzyme catalyzes may not be the same as under artificial conditions. This can result in the same enzyme being identified with two different names. *E.g.* Glucose isomerase, used industrially to convert glucose into the sweetener fructose, is a xylose isomerase *in vivo*.

The International Union of Biochemistry and Molecular Biology have developed a nomenclature for enzymes, the EC numbers; each enzyme is described by a sequence of four numbers preceded by "EC". The first number broadly classifies the enzyme based on its mechanism.

The top-level classification is:

- EC 1 *Oxidoreductases*: catalyse oxidation/reduction reactions
- EC 2 *Transferases*: transfer a functional group (*e.g.* a methyl or phosphate group)
- EC 3 *Hydrolases*: catalyse the hydrolysis of various bonds
- EC 4 *Lyases*: cleave various bonds by means other than hydrolysis and oxidation
- EC 5 *Isomerases*: catalyse isomerization changes within a single molecule
- EC 6 *Ligases*: join two molecules with covalent bonds.

According to the naming conventions, enzymes are generally classified into six main family classes and many sub-family classes. Some web-servers, e.g., EzyPred and bioinformatics tools have been developed to predict which main family class and sub-family class an enzyme molecule belongs to according to its sequence information alone via the pseudo amino acid composition.

Industrial Applications

Enzymes are used in the chemical industry and other industrial applications when extremely specific catalysts are required. However, enzymes in general are limited in the number of reactions they have evolved to catalyse and also by their lack of stability in organic solvents and at high temperatures. Consequently, protein engineering is an active area of research and involves attempts to create new enzymes with novel properties, either through rational design or *in vitro* evolution. These efforts have begun to be successful, and a few enzymes have now been desiged "from scratch" to catalyse reactions that do not occur in nature.

Chapter 2

Understanding DNA and RNA

Nucleic Acid

Nucleic acids are biological molecules essential for life, and include DNA (deoxyribonucleic acid) and RNA (ribonucleic acid). Together with proteins, nucleic acids make up the most important macromolecules; each is found in abundance in all living things. Nucleic acids were first discovered by Friedrich Miescher in 1871. Experimental studies of nucleic acids constitute a major part of modern biological and medical research, and form a foundation for genome and forensic science, as well as the biotechnology and pharmaceutical industries.

Occurrence and Nomenclature

The term *nucleic acid* is the over all name for DNA and RNA, members of a family of biopolymers, and is synonymous with *polynucleotide*. Nucleic acids were named for their initial discovery within the cell nucleus, and for the presence of phosphate groups (related to phosphoric acid).

Although first discovered within the nucleus of eukaryotic cells, nucleic acids are now known to be found in all life forms, including within bacteria, archaea, mitochondria, chloroplasts, viruses and viroids. All living cells and organelles contain both DNA and RNA, while viruses contain either DNA or RNA, but not usually both.

The basic component of biological nucleic acids is the nucleotide, each of which contains a pentose sugar (ribose or deoxyribose), a phosphate group, and a nucleobase. Nucleic acids are also generated within the laboratory, through the use of enzymes (DNA and RNA polymerases) and by solid-phase chemical synthesis. The chemical methods also enable the generation of altered nucleic acids that are not found in nature, for example peptide nucleic acids.

Molecular Composition and Size

Nucleic acids can vary in size, but are generally very large molecules. Indeed, DNA molecules are probably the largest individual molecules known. Well-studied biological nucleic acid molecules range in size from 21 nucleotides (small interfering RNA) to large chromosomes (human chromosome 1 is a single molecule that contains 247 million base pairs). In most cases, naturally occurring DNA molecules are double-stranded and RNA molecules are single-stranded. There are numerous exceptions, however—some viruses have genomes made of double-stranded RNA and other viruses have single-stranded DNA genomes, and, in some circumstances, nucleic acid structures with three or four strands can form.

Nucleic acids are linear polymers (chains) of nucleotides. Each nucleotide consists of three components: a purine or pyrimidine nucleobase (sometimes termed *nitrogenous base* or simply *base*), a pentose sugar; and a phosphate group. The substructure consisting of a nucleobase plus sugar is termed a nucleoside. Nucleic acid types differ in the structure of the sugar in their nucleotides - DNA contains 2'-deoxyribose while RNA contains ribose (where the only difference is the presence of a hydroxyl group). Also, the nucleobases found in the two nucleic acid types are different: adenine, cytosine, and guanine are found in both RNA and DNA, while thymine occurs in DNA and uracil occurs in RNA. The sugars and phosphates in nucleic acids are connected to each other in an alternating chain (sugar-phosphate backbone) through phosphodiester linkages. In conventional nomenclature, the carbons to which the phosphate groups attach are the 3'-end and the 5'-end carbons of the sugar. This gives nucleic acids directionality, and the ends of nucleic acid molecules are referred to as 5'-end and 3'-end. The nucleobases are joined to the sugars via an N-glycosidic linkage involving a nucleobase ring nitrogen (N-1 for pyrimidines and N-9 for purines) and the 1' carbon of the pentose sugar ring. Non-standard nucleosides are also found in both RNA and DNA and usually arise from modification of the standard nucleosides within the DNA molecule or the primary (initial) RNA transcript. Transfer RNA (tRNA) molecules contain a particularly large number of modified nucleosides.

Topology

Double-stranded nucleic acids are made up of complementary sequences, in which extensive Watson-Crick base pairing results in the formation of a highly repeated and quite uniform double-helical three-dimensional structure.

In contrast, single-stranded RNA and DNA molecules are not constrained to a regular double helix, and can adopt highly complex three-dimensional structures that are based on short stretches of intramolecular base-paired sequences that include both Watson-Crick and noncanonical base pairs, as well as a wide range of complex tertiary interactions. Nucleic acid molecules are usually unbranched, and may occur as linear and circular molecules. For example, bacterial chromosomes, plasmids, mitochondrial DNA and chloroplast DNA are usually circular double-stranded DNA molecules, while chromosomes of the eukaryotic nucleus are usually linear double-stranded DNA molecules. Most RNA molecules are linear, single-stranded molecules, but both circular and branched molecules can result from RNA splicing reactions.

Nucleic Acid Sequences

One DNA or RNA molecule differs from another primarily in the sequence of nucleotides. Nucleotide sequences are of great importance in biology, since they carry the ultimate instructions that encode all biological molecules, molecular assemblies, subcellular and cellular structures, organs and organisms, and directly enable cognition, memory and behaviour. Enormous efforts have gone into the development of experimental methods to determine the nucleotide sequence of biological DNA and RNA molecules, and today hundreds of millions of nucleotides are sequenced daily at genome centres and smaller laboratories worldwide.

Types of Nucleic Acids

Deoxyribonucleic Acid: Deoxyribonucleic acid is a nucleic acid that contains the genetic instructions used in the development and functioning of all known living organisms. The main role of DNA molecules is the long-term storage of information and DNA is often compared to a set of blueprints, since it contains the instructions needed to construct other components of cells, such as proteins and RNA molecules. The DNA segments that carry this genetic information are called genes, but other DNA sequences have structural purposes, or are involved in regulating the use of this genetic information.

Ribonucleic Acid: Ribonucleic acid (RNA) functions in converting genetic information from genes into the amino acid sequences of proteins. The three universal types of RNA include transfer RNA (tRNA), messenger RNA (mRNA), and ribosomal RNA (rRNA). Messenger RNA acts to carry genetic sequence information between DNA and ribosomes, directing protein synthesis. Ribosomal RNA is a major component of the ribosome, and catalyzes peptide bond formation. Transfer RNA serves as the carrier molecule for amino acids to be used in protein synthesis, and is responsible for decoding the mRNA. In addition, many other classes of RNA are now known.

Artificial Nucleic Acid Analogs: Artificial nucleic acid analogs have been designed and synthesized by chemists, and include peptide nucleic acid, morpholino- and locked nucleic acid, as well as glycol nucleic acid and threose nucleic acid. Each of these is distinguished from naturally-occurring DNA or RNA by changes to the backbone of the molecule.

DNA

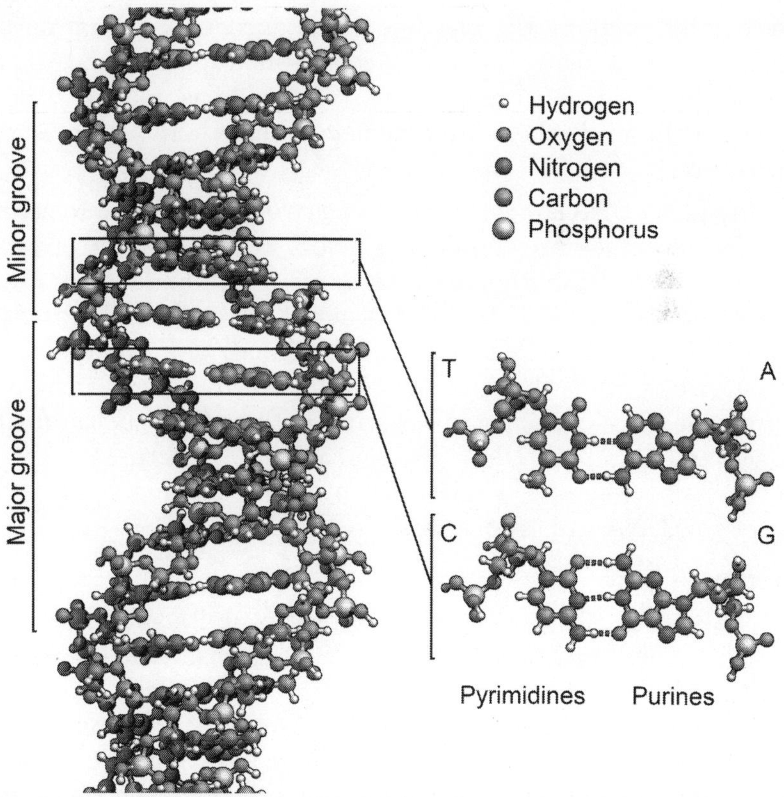

Figure 1: The structure of the DNA double helix. The atoms in the structure are colour coded by element and the detailed structure of two base pairs is shown in the bottom right.

Deoxyribonucleic acid or DNA, is a nucleic acid that contains the genetic instructions used in the development and functioning of all known living organisms (with the exception of RNA viruses). The main role of DNA molecules is the long-term storage of information. DNA is often compared to a set of blueprints, like a recipe or a code, since it contains the instructions needed to construct other components of cells, such as proteins and RNA molecules. The DNA segments that carry this genetic information are called genes, but other DNA sequences have structural purposes, or are involved in regulating the use of this genetic information.

DNA consists of two long polymers of simple units called nucleotides, with backbones made of sugars and phosphate groups joined by ester bonds. These two strands run in opposite directions to each other and are therefore anti-parallel. Attached to each sugar is one of four types of molecules called bases. It is the sequence of these four bases along the backbone that encodes information. This information is read using the genetic code, which specifies the sequence of the amino acids within proteins. The code is read by copying stretches of DNA into the related nucleic acid RNA, in a process called transcription.

Within cells, DNA is organized into long structures called chromosomes. These chromosomes are duplicated before cells divide, in a process called DNA replication. Eukaryotic organisms (animals, plants, fungi, and protists) store most of their DNA inside the cell nucleus and some of their DNA in organelles, such as mitochondria or chloroplasts. In contrast, prokaryotes (bacteria and archaea) store their DNA nly in the cytplasm Within the chromosoes, chromatin prteins such as histones compact and organize DNA. These compact structures guide the interactions between DNA and other proteins, helping control which parts of the DNA are transcribed.

Properties

DNA is a long polymer made from repeating units called nucleotides. As first discovered by James D. Watson and Francis Crick, the structure of DNA of all species comprises two helical chains each coiled round the same axis, and each with a pitch of 34 Ångströms (3.4 nanometres) and a radius of 10 Ångströms (1.0 nanometres). According to another study, when measured in a particular solution, the DNA chain measured 22 to 26 Ångströms wide (2.2 to 2.6 nanometres), and one nucleotide unit measured 3.3 Å (0.33 nm) long. Although each individual repeating unit is very small, DNA polymers can be very large molecules containing millions of nucleotides. For instance, the largest human chromosome, chromosome number 1, is approximately 220 million base pairs long.

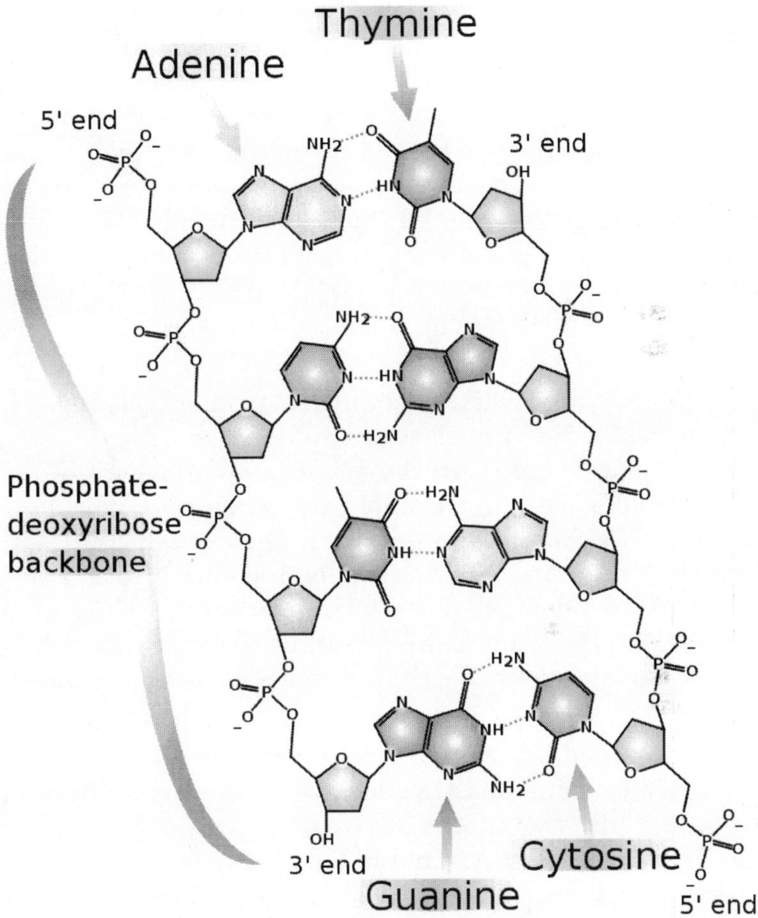

Figure 2: Chemical structure of DNA. Hydrogen bonds shown as dotted lines.

In living organisms, DNA does not usually exist as a single molecule, but instead as a pair of molecules that are held tightly together. These two long strands entwine like vines, in the shape of a double helix. The nucleotide repeats contain both the segment of the backbone of the molecule, which holds the chain together, and a base, which interacts with the other DNA strand in the helix. A base linked to a sugar is called a nucleoside and a base linked to a sugar and one or more phosphae groups is called a nucleotide. If multiple nucleotides are linked together, as in DNA, this polymer is called a polynucleotide. The backbone of the DNA strand is made from alternating phosphate and sugar residues.

The sugar in DNA is 2-deoxyribose, which is a pentose (five-carbon) sugar. The sugars are joined together by phosphate groups that form phosphodiester bonds between the third and fifth carbon atoms of adjacent sugar rings. These asymmetric bonds mean a strand of DNA has a direction. In a double helix the direction of the nucleotides in one strand is opposite to their direction in the other strand: the strands are *antiparallel.* The asymmetric ends of DNA strands are called the 52 (*five prime*) and 32 (*three prime*) ends, with the 5' end having a terminal phosphate group and the 3' end a terminal hydroxyl group. One major difference between DNA and RNA is the sugar, with the 2-deoxyribose in DNA being replaced by the alternative pentose sugar ribose in RNA.

The DNA double helix is stabilized primarily by two forces: hydrogen bonds between nucleotides and base-stacking interactions among the aromatic bases. In the aqueous environment of the cell, the conjugated ∂ bonds of nucleotide bases align perpendicular to the axis of the DNA molecule, minimizing their interaction with the solvation shell and therefore, the Gibbs free energy. The four bases found in DNA are adenine (abbreviated A), cytosine (C), guanine (G) and thymine (T). These four bases are attached to the sugar/phosphate to form the complete nucleotide, as shown for adenosine monophosphate.

These bases are classified into two types; adenine and guanine are fused five- and six-membered heterocyclic compounds called purines, while cytosine and thymine are six-membered rings called pyrimidines. A fifth pyrimidine base, called uracil (U), usually takes the place of thymine in RNA and differs from thymine by lacking a methyl group on its ring. Uracil is not usually found in DNA, occurring only as a breakdown product of cytosine. In addition to RNA and DNA, a large number of artificial nucleic acid analogues have also been created to study the proprieties of nucleic acids, or for use in biotechnology.

Grooves: Twin helical strands form the DNA backbone. Another double helix may be found by tracing the spaces, or grooves, between the strands. These voids are adjacent to the base pairs and may provide a binding site. As the strands are not directly opposite each other, the grooves are unequally sized. One groove, the major groove, is 22 Å wide and the other, the minor groove, is 12 Å wide. The narrowness of the minor groove means that the edges of the bases are more accessible in the major groove. As a result, proteins like transcription factors that can bind to specific sequences in double-stranded DNA usually make contacts to the sides of the bases exposed in the major groove. This situation varies in unusual conformations of DNA within the cell, but the major and minor grooves are always named to reflect the differences in size that would be seen if the DNA is twisted back into the ordinary B form.

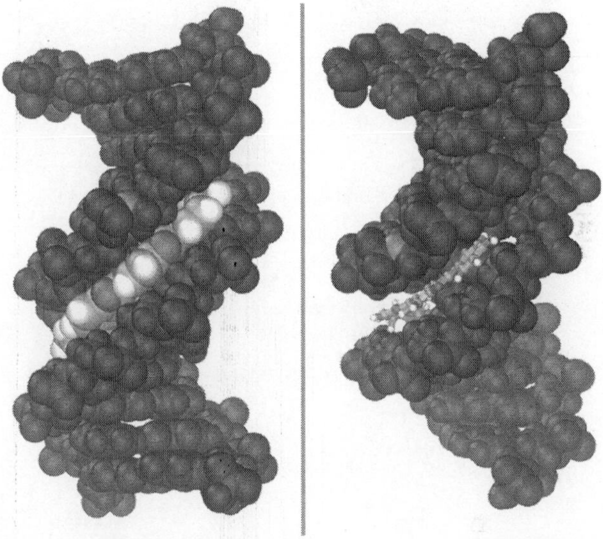

Figure 3: Major and minor grooves of DNA. Minor groove is a binding site for the dye Hoechst 33258.

Base Pairing: Each type of base on one strand forms a bond with just one type of base on the other strand. This is called complementary base pairing. Here, purines form hydrogen bonds to pyrimidines, with A bonding only to T, and C bonding only to G. This arrangement of two nucleotides binding together across the double helix is called a base pair. As hydrogen bonds are not covalent, they can be broken and rejoined relativelyeasily. The two strands of DNA in a double helix can therefore be pulled apart like a zipper, either by a mechanical force or high temperature. As a result of this complementarity, all the information in the double-stranded sequence of a DNA helix is duplicated on each strand, which is vital in DNA replication. Indeed, this reversible and specific interaction between complementary base pairs is critical for all the functions of DNA in living organisms.

The two types of base pairs form different numbers of hydrogen bonds, AT forming two hydrogen bonds, and GC forming three hydrogen bonds. DNA with high GC-content is more stable than DNA with low

GC-content, due t the added stability of an additional hydrogen bond. As a result, it is both the percentage of GC base pairs and the overall length of a DNA double helix that determine the strength of the association between the two strands of DNA.

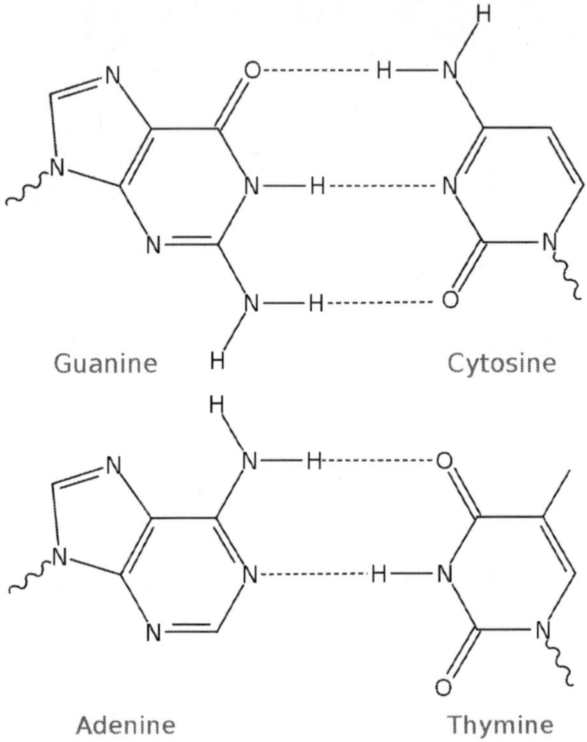

Figure 4: Top, a GC base pair with three hydrogen bonds. Bottom, an AT base pair with two hydrogen bonds. Non-covalent hydrogen bonds between the pairs are shown as dashed lines.

Long DNA helices with a high GC content have stronger-interacting strands, while short helices with high AT content have weaker-interacting strands. In biology, parts of the DNA double helix that need to separate easily, such as the TATAAT Pribnow box in some promoters, tend to have a high AT content, making the strands easier to pull apart. In the laboratory, thestrength of this interaction can be measured by finding the temperature required to break the hydrogen bonds, their melting temperature (also called *Tm* value). When all the base pairs in a DNA double helix melt, the strands separate and exist in solution as two entirely independent molecules. These single-stranded DNA molecules (*ssDNA*) have no single common shape, but some conformations are more stable than others.

Sense and Antisense: A DNA sequence is called "sense" if its sequence is the same as that of a messenger RNA copy that is translated into protein. The sequence on the opposite strand is called the "antisense" sequence. Both sense and antisense sequences can exist on different parts of the same strand of DNA (i.e. both strands contain both sense and antisense sequences). In both prokaryotes and eukaryotes, antisense RNA sequences are produced, but the functions of these RNAs are not entirely clear. One proposal is that antisense RNAs are involved in regulating gene exression through RNA-RNA base pairing.

A few DNA sequences in prokaryotes and eukaryotes, and more in plasmids and viruses, blur the distinction between sense and antisense strands by having overlapping genes. In these cases, some DNA sequences do double duty, encoding one protein when read along one strand, and a second protein when read in the opposite direction along the other strand. In bacteria, this overlap may be involved in the regulation of gene transcription, while in viruses, overlapping genes increase the amount of information that can be encoded within the small viral genome.

Supercoiling: DNA can be twisted like a rope in a process called DNA supercoiling. With DNA in its "relaxed" state, a strand usually circles the axis of the double helix once every 10.4 base pairs, but if the DNA is twisted the strands become more tightly or more loosely wound. If the DNA is twisted in the direction of the helix, this is positive supercoiling, and the bases are held more tightly together. If they are twisted in the opposite direction, this is negative supercoiling, and the bases come apart more easily. In nature, most DNA has slight negative supercoiling that is introduced by enzymes called topoisomerases. These enzymes are also needed to relieve the twisting stresses introduced into DNA strands during processes such as transcription and DNA replication.

Alternate DNA Structures: DNA exists in many possible conformations that include A-DNA, B-DNA, and Z-DNA forms, although, only B-DNA and Z-DNA have been directly observed in functional organisms. The conformation that DNA adopts depends on the hydration level, DNA sequence, the amount and direction of supercoiling, chemical modifications of the bases, the type and concentration of metal ions, as well as the presence of polyamines in solution.

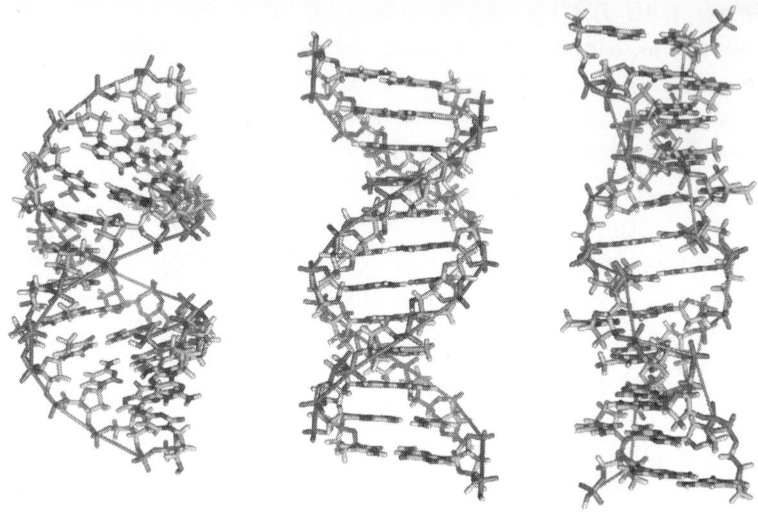

Figure 5: From left to right, the structures of A, B and Z DNA

The first published reports of A-DNA X-ray diffraction patterns—and also B-DNA used analyses based on Patterson transforms that provided only a limited amount of structural information for oriented fibres of DNA. An alternate analysis was then proposed by Wilkins *et al.*, in 1953, for the *in vivo* B-DNA X-ray diffraction/scattering patterns of highly hydrated DNA fibres in terms of squares of Bessel functions. In the same journal, James D. Watson and Francis Crick presented their molecular modelling analysis of the DNA X-ray diffraction patterns to suggest that the structure was a double-helix.

Although the 'B-DNA form' is most common under the conditions found in cells, it is not a well-defined conformation but a family of related DNA conformations that occur at the high hydration levels present in living cells. Their corresponding X-ray diffraction and scattering patterns are characteristic of molecular paracrystals with a significant degree of disorder. Compared to B-DNA, the A-DNA form is a wider right-handed spiral, with a shallow, wide minor groove and a narrower, deeper major groove. The A form occurs under non-physiological conditions in partially dehydrated samples of DNA, while in the cell it may be produced in hybrid pairings of DNA and RNA strands, as well as in enzyme-DNA complexes. Segments of DNA where the bases have been chemically modified by methylation may undergo a larger change in conformation and adopt the Z form. Here, the strands turn about the helical axis in a left-handed spiral, the opposite of the more common B form. These unusual structures can be recognized by specific Z-DNA binding proteins and may be involved in the regulation of transcription.

Alternate DNA Chemistry: For a number of years exobiologists have proposed the existence of a shadow biosphere, a postulated microbial biosphere of Earth that uses radically different biochemical and molecular processes than currently known life. One of the proposals was the existence of lifeforms that use arsenic instead of phosphorus in DNA.

A December 2010 NASA press conference revealed that the bacterium GFAJ-1, which has evolved in an arsenic-rich environment, is the first terrestrial lifeform found which may have this ability. The bacterium was found in Mono Lake, east of Yosemite National Park. GFAJ-1 is a rod-shaped extremophile bacterium in the family Halomonadaceae that, when starved of phosphorus, may be capable of incorporating the usually poisonous element arsenic in its DNA. This discovery may lend weight to the long-standing idea that extraterrestrial life could have a different chemical makeup from life on Earth. The research was carried out by a team led by Felisa Wolfe-Simon, a geomicrobiologist and geobiochemist, a Postdoctoral Fellow of the NASA Astrobiology Institute with Arizona State University. This finding has, however, faced strong criticism from the scientific community; scientists have argued that there is no evidence that arsenic is actually incorporated into biomolecules. Independent conformation of this finding has also not yet been possible.

Quadruplex Structures: At the ends of the linear chromosomes are specialized regions of DNA called telomeres.

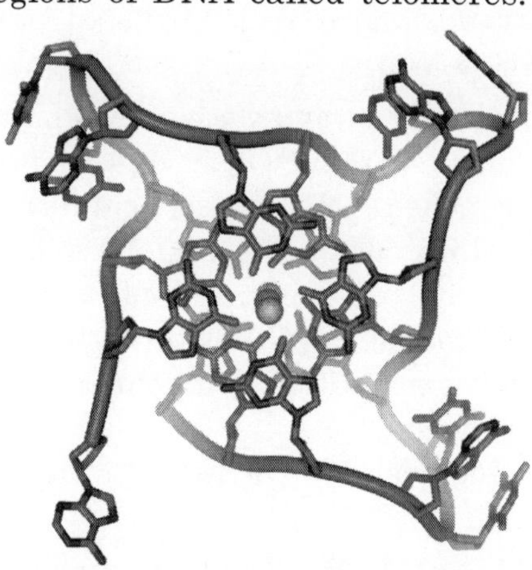

Figure 6: DNA quadruplex formed by telomere repeats. The looped conformation of the DNA backbone is very different from the typical DNA helix.

The main function of these regions is to allow the cell to replicate chromosome ends using the enzyme telomerase, as the enzymes that normally replicate DNA cannot copy the extreme 32 ends of chromosomes. These specialized chromosome caps also help protect the DNA ends, and stop the DNA repair systems in the cell from treating them as damage to be corrected. In human cells, telomeres are usually lengths of single-stranded DNA containing several thousand repeats of a simple TTAGGG sequence.

These guanine-rich sequences may stabilize chromosome ends by forming structures of stacked sets of four-base units, rather than the usual base pairs found in other DNA molecules. Here, four guanine bases form a flat plate and these flat four-base units then stack on top of each other, to form a stable *G-quadruplex* structure. These structures are stabilized by hydrogen bonding between the edges of the bases and chelation of a metal ion in the centre of each four-base unit. Other structures can also be formed, with the central set of four bases coming from either a single strand folded around the bases, or several different parallel strands, each contributing one base to the central structure.

In addition to these stacked structures, telomeres also form large loop structures called telomere loops, or T-loops. Here, the single-stranded DNA curls around in a long circle stabilized by telomere-binding proteins. At the very end of the T-loop, the single-stranded telomere DNA is held onto a region of double-stranded DNA by the telomere strand disrupting the double-helical DNA and base pairing to one of the two strands. This triple-stranded structure is called a displacement loop or D-loop.

Branched DNA: In DNA fraying occurs when non-complementary regions exist at the end of an otherwise complementary double-strand of DNA. However, branched DNA can occur if a third strand of DNA is introduced and contains adjoining regions able to hybridize with the frayed regions of the pre-existing double-strand. Although the simplest example of branched DNA involves only three strands of DNA, complexes involving additional strands and multiple branches are also possible. Branched DNA can be used in nanotechnology to construct geometric shapes, see the section on uses in technology below.

Vibration: DNA may carry out low-frequency collective motion as observed by the Raman spectroscopy and analyzed with a quasi-continuum model.

Chemical Modifications

Base Modifications: The expression of genes is influenced by how the DNA is packaged in chromosomes, in a structure called chromatin. Base modifications can be involved in packaging, with regions that have low or no gene expression usually containing high levels of methylation of cytosine bases. For example, cytosine methylation, produces 5-methylcytosine, which is important for X-chromosome inactivation. The average level of methylation varies between organisms - the worm *Caenorhabditis elegans* lacks cytosine methylation, while vertebrates have higher levels, with up to 1% of their DNA containing 5-methylcytosine. Despite the importance of 5-methylcytosine, it can deaminate to leave a thymine base, so methylated cytosines are particularly prone to mutations. Other base modifications include adenine methylation in bacteria, the presence of 5-hydroxymethylcytosine in the brain, and the glycosylation of uracil to produce the "J-base" in kinetoplastids.

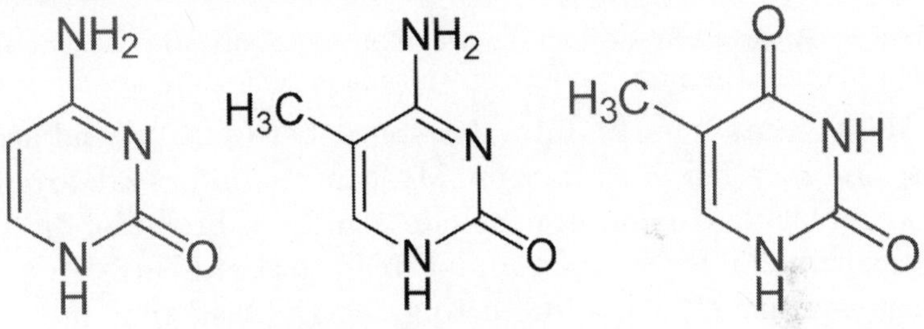

Figure 7: Structure of cytosine with and without the 5-methyl group. Deamination converts 5-methylcytosine into thymine.

Damage: DNA can be damaged by many sorts of mutagens, which change the DNA sequence. Mutagens include oxidizing agents, alkylating agents and also high-energy electromagnetic radiation such as ultraviolet light and X-rays. The type of DNA damage produced depends on the type of mutagen. For example, UV light can damage DNA by producing thymine dimers, which are cross-links between pyrimidine bases. On the other hand, oxidants such as free radicals or hydrogen peroxide produce multiple forms of damage, including base modifications, particularly of guanosine, and double-strand breaks. A typical human cell contains about 150,000 bases that have suffered oxidative damage. Of these oxidative lesions, the most dangerous are double-strand breaks, as these are difficult to repair and can produce point mutations, insertions and deletions from the DNA sequence, as well as chromosomal translocations.

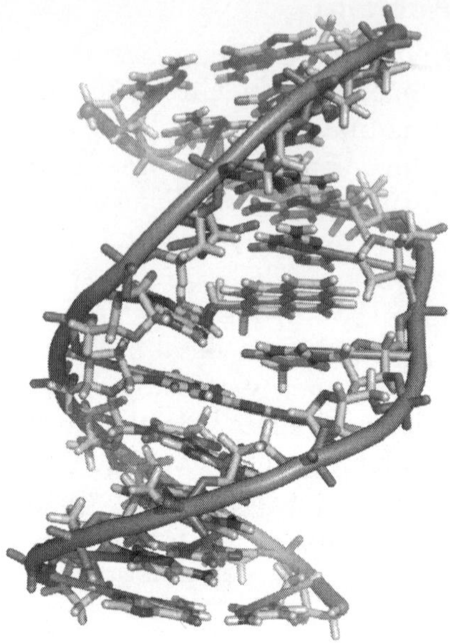

Figure 8: A covalent adduct between a metabolically activated form of benzo[a]pyrene, the major mutagen in tobacco smoke, and DNA

Many mutagens fit into the space between two adjacent base pairs, this is called *intercalation*. Most intercalators are aromatic and planar molecules; examples include ethidium bromide, daunomycin, and doxorubicin. In order for an intercalator to fit between base pairs, the bases must separate, distorting the DNA strands by unwinding of the double helix. This inhibits both transcription and DNA replication, causing toxicity and mutations. As a result, DNA intercalators are often carcinogens, and benzo[*a*]pyrene diol epoxide, acridines, aflatoxin and ethidium bromide are well-known examples. Nevertheless, due to their ability to inhibit DNA transcription and replication, other similar toxins are also used in chemotherapy to inhibit rapidly growing cancer cells.

Biological Functions: DNA usually occurs as linear chromosomes in eukaryotes, and circular chromosomes in prokaryotes. The set of chromosomes in a cell makes up its genome; the human genome has approximately 3 billion base pairs of DNA arranged into 46 chromosomes. The information carried by DNA is held in the sequence of pieces of DNA called genes. Transmission of genetic information in genes is achieved via complementary base pairing. For example, in transcription, when a cell uses the information in a gene, the DNA sequence is copied into a complementary RNA sequence through the attraction between the DNA and the correct RNA nucleotides. Usually, this RNA copy is then used to make a matching protein sequence in a process called translation, which depends on the same interaction between RNA nucleotides. In alternative fashion, a cell may simply copy its genetic information in a process called DNA replication. The details of these functions are covered in other articles; here we focus on the interactions between DNA and other molecules that mediate the function of the genome.

Genes and Genomes: Genomic DNA is tightly and orderly packed in the process called DNA condensation to fit the small available volumes of the cell. In eukaryotes, DNA is located in the cell nucleus, as well as small amounts in mitochondria and chloroplasts. In prokaryotes, the DNA is held within an irregularly shaped body in the cytoplasm called the nucleoid. The genetic information in a genome is held within genes, and the complete set of this information in an organism is called its genotype. A gene is a unit of heredity and is a region of DNA that influences a particular characteristic in an organism. Genes contain an open reading frame that can be transcribed, as well as regulatory sequences such as promoters and enhancers, which control the transcription of the open reading frame.

In many species, only a small fraction of the total sequence of the genome encodes protein. For example, only about 1.5% of the human genome consists of protein-coding exons, with over 50% of human DNA consisting of non-coding repetitive sequences. The reasons for the presence of so much noncoding DNA in eukaryotic genomes and the extraordinary differences in genome size, or *C-value*, among species represent a long-standing puzzle known as the "C-value enigma". However, DNA sequences that do not code protein may still encode functional non-coding RNA molecules, which are involved in the regulation of gene expression. Some noncoding DNA sequences play structural roles in chromosomes. Telomeres and centromeres typically contain few genes, but are important for the function and stability of chromosomes. An abundant form of noncoding DNA in humans are pseudogenes, which are copies of genes that have been disabled by mutation. These sequences are usually just molecular fossils, although they can occasionally serve as raw genetic material for the creation of new genes through the process of gene duplication and divergence.

Transcription and Translation: A gene is a sequence of DNA that contains genetic information and can influence the phenotype of an organism. Within a gene, the sequence of bases along a DNA strand defines a messenger RNA sequence, which then defines one or more protein sequences. The relationship between the nucleotide sequences of genes and the amino-acid sequences of proteins is determined by the rules of translation, known collectively as the genetic code. The genetic code consists of three-letter 'words' called *codons* formed from a sequence of three nucleotides (e.g. ACT, CAG, TTT). In transcription, the codons of a gene are copied into messenger RNA by RNA polymerase. This RNA copy is then decoded by a ribosome that reads the RNA sequence by base-pairing the messenger RNA to transfer RNA, which carries amino acids. Since there are 4 bases in 3-letter combinations, there are 64 possible codons (43 combinations). These encode the twenty standard amino acids, giving most amino acids more than one possible codon. There are also three 'stop' or 'nonsense' codons signifying the end of the coding region; these are the TAA, TGA and TAG codons.

Replication: Cell division is essential for an organism to grow, but, when a cell divides, it must replicate the DNA in its genome so that the two daughter cells have the same genetic information as their parent.

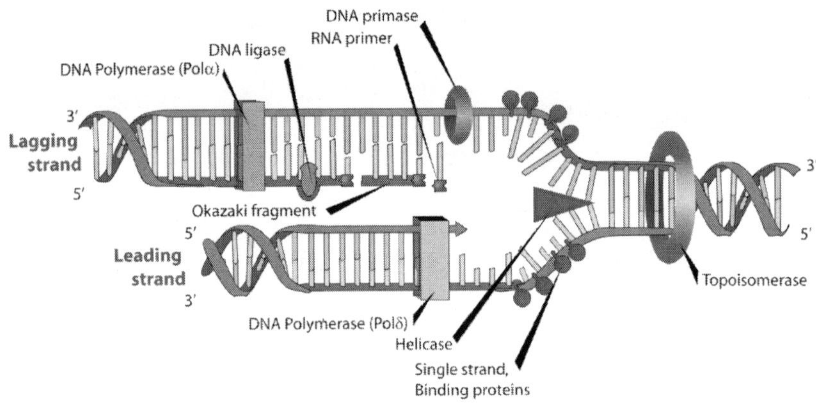

Figure 9: DNA replication. The double helix is unwound by a helicase and topoisomerase. Next, one DNA polymerase produces the leading strand copy. Another DNA polymerase binds to the lagging strand. This enzyme makes discontinuous segments (called Okazaki fragments) before DNA ligase joins them together.

The double-stranded structure of DNA provides a simple mechanism for DNA replication. Here, the two strands are separated and then each strand's complementary DNA sequence is recreated by an enzyme called DNA polymerase. This enzyme makes the complementary strand by finding the correct base through complementary base pairing, and bonding it onto the original strand. As DNA polymerases can only extend a DNA strand in a 52 to 32 direction, different mechanisms are used to copy the antiparallel strands of the double helix. In this way, the base on the old strand dictates which base appears on the new strand, and the cell ends up with a perfect copy of its DNA.

Interactions with Proteins: All the functions of DNA depend on interactions with proteins. These protein interactions can be non-specific, or the protein can bind specifically to a single DNA sequence. Enzymes can also bind to DNA and of these, the polymerases that copy the DNA base sequence in transcription and DNA replication are particularly important.

DNA-Binding Proteins: Structural proteins that bind DNA are well-understood examples of non-specific DNA-protein interactions. Within chromosomes, DNA is held in complexes with structural proteins. These proteins organize the DNA into a compact structure called chromatin. In eukaryotes this structure involves DNA binding to a complex of small basic proteins called histones, while in prokaryotes multiple types of proteins are involved. The histones form a disk-shaped complex called a nucleosome, which contains two complete turns of double-stranded DNA wrapped around its surface. These non-specific interactions are formed through basic residues in the histones making ionic bonds to the acidic sugar-phosphate backbone of the DNA, and are therefore largely independent of the base sequence.

Chemical modifications of these basic amino acid residues include methylation, phosphorylation and acetylation. These chemical changes alter the strength of the interaction between the DNA and the histones, making the DNA more or less accessible to transcription factors and changing the rate of transcription. Other non-specific DNA-binding proteins in chromatin include the high-mobility group proteins, which bind to bent or distorted DNA. These proteins are important in bending arrays of nucleosomes and arranging them into the larger structures that make up chromosomes. A distinct group of DNA-binding proteins are the DNA-binding proteins that specifically bind single-stranded DNA.

In humans, replication protein A is the best-understood member of this family and is used in processes where the double helix is separated, including DNA replication, recombination and DNA repair. These binding proteins seem to stabilize single-stranded DNA and protect it from forming stem-loops or being degraded by nucleases. In contrast, other proteins have evolved to bind to particular DNA sequences. The most intensively studied of these are the various transcription factors, which are proteins that regulate transcription. Each transcription factor binds to one particular set of DNA sequences and activates or inhibits the transcription of genes that have these sequences close to their promoters.

The transcription factors do this in two ways. Firstly, they can bind the RNA polymerase responsible for transcription, either directly or through other mediator proteins; this locates the polymerase at the promoter and allows it to begin transcription. Alternatively, transcription factors can bind enzymes that modify the histones at the promoter; this will change the accessibility of the DNA template to the polymerase. As these DNA targets can occur throughout an organism's genome, changes in the activity of one type of transcription factor can affect thousands of genes.

Consequently, these proteins are often the targets of the signal transduction processes that control responses to environmental changes or cellular differentiation and development. The specificity of these transcription factors' interactions with DNA come from the proteins making multiple contacts to the edges of the DNA bases, allowing them to "read" the DNA sequence. Most of these base-interactions are made in the major groove, where the bases are most accessible.

DNA-Modifying Enzymes

Nucleases and Ligases: Nucleases are enzymes that cut DNA strands by catalyzing the hydrolysis of the phosphodiester bonds. Nucleases that hydrolyse nucleotides from the ends of DNA strands are called exonucleases, while endonucleases cut within strands. The most frequently used nucleases in molecular biology are the restriction endonucleases, which cut DNA at specific sequences. For instance, the EcoRV enzyme shown to the left recognizes the 6-base sequence 52 -GAT | ATC-32 and makes a cut at the vertical line. In nature, these enzymes protect bacteria against phage infection by digesting the phage DNA when it enters the bacterial cell, acting as part of the restriction modification system. In technology, these sequence-specific nucleases are used in molecular cloning and DNA fingerprinting.

Enzymes called DNA ligases can rejoin cut or broken DNA strands. Ligases are particularly important in lagging strand DNA replication, as they join together the short segments of DNA produced at the replication fork into a complete copy of the DNA template. They are also used in DNA repair and genetic recombination.

Topoisomerases and Helicases: Topoisomerases are enzymes with both nuclease and ligase activity. These proteins change the amount of supercoiling in DNA. Some of these enzymes work by cutting the DNA helix and allowing one section to rotate, thereby reducing its level of supercoiling; the enzyme then seals the DNA break. Other types of these enzymes are capable of cutting one DNA helix and then passing a second strand of DNA through this break, before rejoining the helix. Topoisomerases are required for many processes involving DNA, such as DNA replication and transcription. Helicases are proteins that are a type of molecular motor. They use the chemical energy in nucleoside triphosphates, predominantly ATP, to break hydrogen bonds between bases and unwind the DNA double helix into single strands. These enzymes are essential for most processes where enzymes need to access the DNA bases.

Polymerases: Polymerases are enzymes that synthesize polynucleotide chains from nucleoside triphosphates. The sequence of their products are copies of existing polynucleotide chains - which are called *templates*. These enzymes function by adding nucleotides onto the 32 hydroxyl group of the previous nucleotide in a DNA strand. As a consequence, all polymerases work in a 52 to 32 direction. In the active site of these enzymes, the incoming nucleoside triphosphate base-pairs to the template: this allows polymerases to accurately synthesize the complementary strand of their template. Polymerases are classified according to the type of template that they use.

In DNA replication, a DNA-dependent DNA polymerase makes a copy of a DNA sequence. Accuracy is vital in this process, so many of these polymerases have a proofreading activity. Here, the polymerase recognizes the occasional mistakes in the synthesis reaction by the lack of base pairing between the mismatched nucleotides. If a mismatch is detected, a 32 to 52 exonuclease activity is activated and the incorrect base removed. In most organisms, DNA polymerases function in a large complex called the replisome that contains multiple accessory subunits, such as the DNA clamp or helicases.

RNA-dependent DNA polymerases are a specialized class of polymerases that copy the sequence of an RNA strand into DNA. They include reverse transcriptase, which is a viral enzyme involved in the infection of cells by retroviruses, and telomerase, which is required for the replication of telomeres. Telomerase is an unusual polymerase because it contains its own RNA template as part of its structure.

Transcription is carried out by a DNA-dependent RNA polymerase that copies the sequence of a DNA strand into RNA. To begin transcribing a gene, the RNA polymerase binds to a sequence of DNA called a promoter and separates the DNA strands. It then copies the gene sequence into a messenger RNA transcript until it reaches a region of DNA called the terminator, where it halts and detaches from the DNA. As with human DNA-dependent DNA polymerases, RNA polymerase II, the enzyme that transcribes most of the genes in the human genome, operates as part of a large protein complex with multiple regulatory and accessory subunits.

Genetic Recombination

DNA helix usually does not interact with other segments of DNA, and in human cells the different chromosomes even occupy separate areas in the nucleus called "chromosome territories". This physical separation of different chromosomes is important for the ability of DNA to function as a stable repository for information, as one of the few times chromosomes interact is during chromosomal crossover when they recombine. Chromosomal crossover is when two DNA helices break, swap a section and then rejoin.

Recombination allows chromosomes to exchange genetic information and produces new combinations of genes, which increases the efficiency of natural selection and can be important in the rapid evolution of new proteins. Genetic recombination can also be involved in DNA repair, particularly in the cell's response to double-strand breaks. The most common form of chromosomal crossover is homologous recombination, where the two chromosomes involved share very similar sequences. Non-homologous recombination can be damaging to cells, as it can produce chromosomal translocations and genetic abnormalities.

The recombination reaction is catalyzed by enzymes known as recombinases, such as RAD51. The first step in recombination is a double-stranded break either caused by an endonuclease or damage to the DNA. A series of steps catalyzed in part by the recombinase then leads to joining of the two helices by at least one Holliday junction, in which a segment of a single strand in each helix is annealed to the complementary strand in the other helix. The Holliday junction is a tetrahedral junction structure that can be moved along the pair of chromosomes, swapping one strand for another. The recombination reaction is then halted by cleavage of the junction and re-ligation of the released DNA.

Evolution

DNA contains the genetic information that allows all modern living things to function, grow and reproduce. However, it is unclear how long in the 4-billion-year history of life DNA has performed this function, as it has been proposed that the earliest forms of life may have used RNA as their genetic material. RNA may have acted as the central part of early cell metabolism as it can both transmit genetic information and carry out catalysis as part of ribozymes. This ancient RNA world where nucleic acid would have been used for both catalysis and genetics may have influenced the evolution of the current genetic code based on four nucleotide bases.

This would occur, since the number of different bases in such an organism is a trade-off between a small number of bases increasing replication accuracy and a large number of bases increasing the catalytic efficiency of ribozymes. However, there is no direct evidence of ancient genetic systems, as recovery of DNA from most fossils is impossible. This is because DNA will survive in the environment for less than one million years and slowly degrades into short fragments in solution. Claims for older DNA have been made, most notably a report of the isolation of a viable bacterium from a salt crystal 250 million years old, but these claims are controversial.

Uses in Technology

Genetic Engineering: Methods have been developed to purify DNA from organisms, such as phenol-chloroform extraction, and to manipulate it in the laboratory, such as restriction digests and the polymerase chain reaction. Modern biology and biochemistry make intensive use of these techniques in recombinant DNA technology. Recombinant DNA is a man-made DNA sequence that has been assembled from other DNA sequences. They can be transformed into organisms in the form of plasmids or in the appropriate format, by using a viral vector. The genetically modified organisms produced can be used to produce products such as recombinant proteins, used in medical research, or be grown in agriculture.

Forensics: Forensic scientists can use DNA in blood, semen, skin, saliva or hair found at a crime scene to identify a matching DNA of an individual, such as a perpetrator. This process is formally termed DNA profiling, but may also be called "genetic fingerprinting". In DNA profiling, the lengths of variable sections of repetitive DNA, such as short tandem repeats and minisatellites, are compared between people. This method is usually an extremely reliable technique for identifying a matching DNA. However, identification can be complicated if the scene is contaminated with DNA from several people. DNA profiling was developed in 1984 by British geneticist Sir Alec Jeffreys, and first used in forensic science to convict Colin Pitchfork in the 1988 Enderby murders case.

People convicted of certain types of crimes may be required to provide a sample of DNA for a database. This has helped investigators solve old cases where only a DNA sample was obtained from the scene. DNA profiling can also be used to identify victims of mass casualty incidents. On the other hand, many convicted people have been released from prison on the basis of DNA techniques, which were not available when a crime had originally been committed.

Bioinformatics: Bioinformatics involves the manipulation, searching, and data mining of biological data, and this includes DNA sequence data. The development of techniques to store and search DNA sequences have led to widely applied advances in computer science, especially string searching algorithms, machine learning and database theory. String searching or matching algorithms, which find an occurrence of a sequence of letters inside a larger sequence of letters, were developed to search for specific sequences of nucleotides. The DNA sequenced may be aligned with other DNA sequences to identify homologous sequences and locate the specific mutations that make them distinct.

These techniques, especially multiple sequence alignment, are used in studying phylogenetic relationships and protein function. Data sets representing entire genomes' worth of DNA sequences, such as those produced by the Human Genome Project, are difficult to use without the annotations that identify the locations of genes and regulatory elements on each chromosome. Regions of DNA sequence that have the characteristic patterns associated with protein- or RNA-coding genes can be identified by gene finding algorithms, which allow researchers to predict the presence of particular gene products and their possible functions in an organism even before they have been isolated experimentally. Entire genomes may also be compared which can shed light on the evolutionary history of particular organism and permit the examination of complex evolutionary events.

DNA Nanotechnology: DNA nanotechnology uses the unique molecular recognition properties of DNA and other nucleic acids to create self-assembling branched DNA complexes with useful properties. DNA is thus used as a structural material rather than as a carrier of biological information. This has led to the creation of two-dimensional periodic lattices (both tile-based as well as using the "DNA origami" method) as well as three-dimensional structures in the shapes of polyhedra. Nanomechanical devices and algorithmic self-assembly have also been demonstrated, and these DNA structures have been used to template the arrangement of other molecules such as gold nanoparticles and streptavidin proteins.

History and Anthropology: Because DNA collects mutations over time, which are then inherited, it contains historical information, and, by comparing DNA sequences, geneticists can infer the evolutionary history of organisms, their phylogeny. This field of phylogenetics is a powerful tool in evolutionary biology. If DNA sequences within a species are compared, population geneticists can learn the history of particular populations. This can be used in studies ranging from ecological genetics to anthropology; For example, DNA evidence is being used to try to identify the Ten Lost Tribes of Israel. DNA has also been used to look at modern family relationships, such as establishing family relationships between the descendants of Sally Hemings and Thomas Jefferson. This usage is closely related to the use of DNA in criminal investigations detailed above. Indeed, some criminal investigations have been solved when DNA from crime scenes has matched relatives of the guilty individual.

History of DNA Research

DNA was first isolated by the Swiss physician Friedrich Miescher who, in 1869, discovered a microscopic substance in the pus of discarded surgical bandages. As it resided in the nuclei of cells, he called it "nuclein". In 1919, Phoebus Levene identified the base, sugar and phosphate nucleotide unit. Levene suggested that DNA consisted of a string of nucleotide units linked together through the phosphate groups. However, Levene thought the chain was short and the bases repeated in a fixed order. In 1937 William Astbury produced the first X-ray diffraction patterns that showed that DNA had a regular structure.

In 1928, Frederick Griffith discovered that traits of the "smooth" form of the *Pneumococcus* could be transferred to the "rough" form of the same bacteria by mixing killed "smooth" bacteria with the live "rough" form. This system provided the first clear suggestion that DNA carries genetic information—the Avery–MacLeod–McCarty experiment—when Oswald Avery, along with coworkers Colin MacLeod and Maclyn McCarty, identified DNA as the transforming principle in 1943. DNA's role in heredity was confirmed in 1952, when Alfred Hershey and Martha Chase in the Hershey–Chase experiment showed that DNA is the genetic material of the T2 phage.

In 1953, James D. Watson and Francis Crick suggested what is now accepted as the first correct double-helix model of DNA structure in the journal *Nature*. Their double-helix, molecular model of DNA was then based on a single X-ray diffraction image (labeled as "Photo 51") taken by Rosalind Franklin and Raymond Gosling in May 1952, as well as the information that the DNA bases are paired — also obtained through private communications from Erwin Chargaff in the previous years. Chargaff's rules played a very important role in establishing double-helix configurations for B-DNA as well as A-DNA.

Experimental evidence supporting the Watson and Crick model were published in a series of five articles in the same issue of *Nature*. Of these, Franklin and Gosling's paper was the first publication of their own X-ray diffraction data and original analysis method that partially supported the Watson and Crick model; this issue also contained an article on DNA structure by Maurice Wilkins and two of his colleagues, whose analysis and *in vivo* B-DNA X-ray patterns also supported the presence *in vivo* of the double-helical DNA configurations as proposed by Crick and Watson for their double-helix molecular model of DNA in the previous two pages of *Nature*. In 1962, after Franklin's death, Watson, Crick, and Wilkins jointly received the Nobel Prize in Physiology or Medicine. However, Nobel rules of the time allowed only living recipients, but a vigorous debate continues on who should receive credit for the discovery.

In an influential presentation in 1957, Crick laid out the central dogma of molecular biology, which foretold the relationship between DNA, RNA, and proteins, and articulated the "adaptor hypothesis". Final confirmation of the replication mechanism that was implied by the double-helical structure followed in 1958 through the Meselson–Stahl experiment. Further work by Crick and coworkers showed that the genetic code was based on non-overlapping triplets of bases, called codons, allowing Har Gobind Khorana, Robert W. Holley and Marshall Warren Nirenberg to decipher the genetic code. These findings represent the birth of molecular biology.

RNA

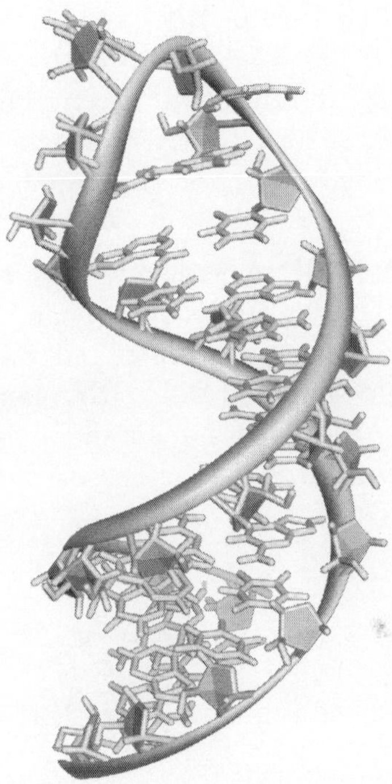

Figure 10: A hairpin loop from a pre-mRNA. Highlighted are the nucleobases (green) and the ribose-phosphate backbone

Ribonucleic acid (RNA) is one of the three major macromolecules (along with DNA and proteins) that are essential for all known forms of life. Like DNA, RNA is made up of a long chain of components called nucleotides. Each nucleotide consists of a nucleobase (sometimes called a nitrogenous base), a ribose sugar, and a phosphate group. The sequence of nucleotides allows RNA to encode genetic information. For example, some viruses use RNA instead of DNA as their genetic material, and all organisms use messenger RNA (mRNA) to carry the genetic information that directs the synthesis of proteins.

Like proteins, some RNA molecules play an active role in cells by catalyzing biological reactions, controlling gene expression, or sensing and communicating responses to cellular signals. One of these active processes is protein synthesis, a universal function whereby mRNA molecules direct the assembly of proteins on ribosomes. This process uses transfer RNA (tRNA) molecules to deliver amino acids to the ribosome, where ribosomal RNA (rRNA) links amino acids together to form proteins.

The chemical structure of RNA is very similar to that of DNA, with two differences—(a) RNA contains the sugar ribose while DNA contains the slightly different sugar deoxyribose (a type of ribose that lacks one oxygen atom), and (b) RNA has the nucleobase uracil while DNA contains thymine (uracil and thymine have similar base-pairing properties).

Unlike DNA, most RNA molecules are single-stranded. Single-stranded RNA molecules adopt very complex three-dimensional structures, since they are not restricted to the repetitive double-helical form of double-stranded DNA. RNA is made within living cells by RNA polymerases, enzymes that act to copy a DNA or RNA template into a new RNA strand through processes known as transcription or RNA replication, respectively.

Comparison with DNA

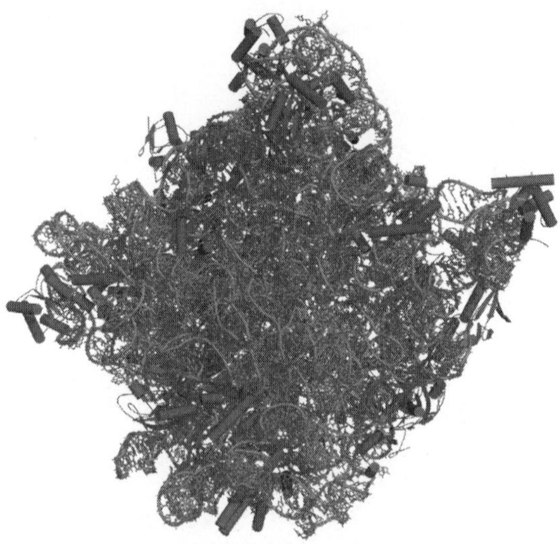

Figure 11: Three-dimensional representation of the 50S ribosomal subunit. RNA is in ochre, protein in blue. The active site is in the middle

RNA and DNA are both nucleic acids, but differ in three main ways. First, unlike DNA, which is, in general, double-stranded, RNA is a single-stranded molecule in many of its biological roles and has a much shorter chain of nucleotides. Second, while DNA contains *deoxyribose*, RNA contains *ribose* (in deoxyribose there is no hydroxyl group attached to the pentose ring in the 2' position). These hydroxyl groups make RNA less stable than DNA because it is more prone to hydrolysis. Third, the complementary base to adenine is not thymine, as it is in DNA, but rather uracil, which is an unmethylated form of thymine.

Like DNA, most biologically active RNAs, including mRNA, tRNA, rRNA, snRNAs, and other non-coding RNAs, contain self-complementary sequences that allow parts of the RNA to fold and pair with itself to form double helices. Structural analysis of these RNAs has revealed that they are highly structured. Unlike DNA, their structures do not consist of long double helices but rather collections of short helices packed together into structures akin to proteins. In this fashion, RNAs can achieve chemical catalysis, like enzymes. For instance, determination of the structure of the ribosome—an enzyme that catalyzes peptide bond formation—revealed that its active site is composed entirely of RNA.

Structure

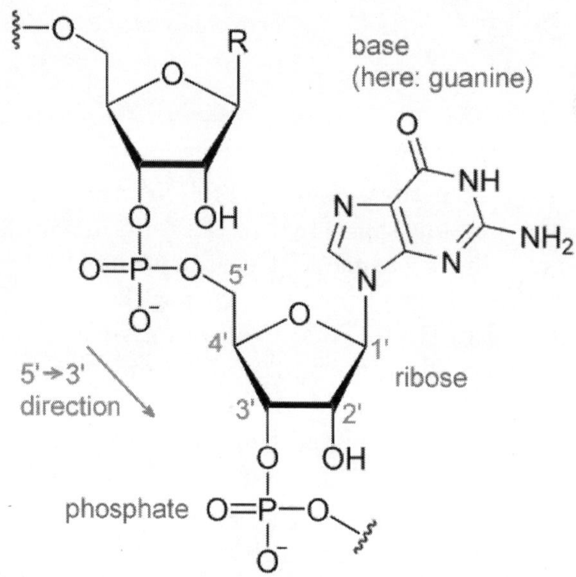

Figure 12: Chemical structure of RNA

Each nucleotide in RNA contains a ribose sugar, with carbons numbered 1' through 5'. A base is attached to the 1' position, in general, adenine (A), cytosine (C), guanine (G), or uracil (U). Adenine and guanine are purines, cytosine, and uracil are pyrimidines. A phosphate group is attached to the 3' position of one ribose and the 5' position of the next. The phosphate groups have a negative charge each at physiological pH, making RNA a charged molecule (polyanion). The bases may form hydrogen bonds between cytosine and guanine, between adenine and uracil and between guanine and uracil. However, other interactions are possible, such as a group of adenine bases binding to each other in a bulge, or the GNRA tetraloop that has a guanine–adenine base-pair.

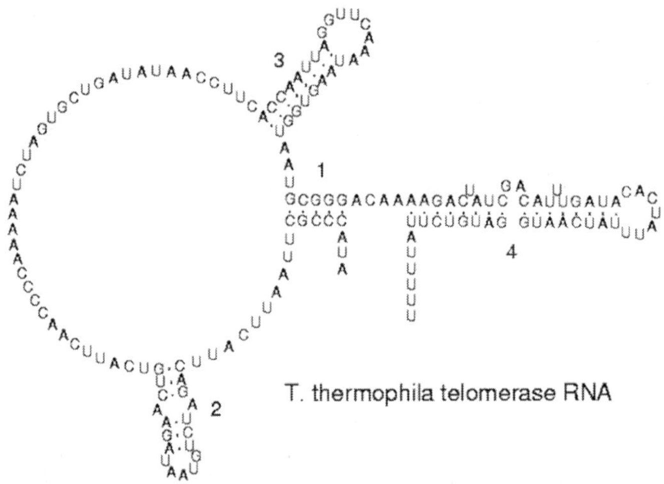

Figure 13: Secondary structure of a telomerase RNA.

An important structural feature of RNA that distinguishes it from DNA is the presence of a hydroxyl group at the 2' position of the ribose sugar. The presence of this functional group causes the helix to adopt the A-form geometry rather than the B-form most commonly observed in DNA. This results in a very deep and narrow major groove and a shallow and wide minor groove. A second consequence of the presence of the 2'-hydroxyl group is that in conformationally flexible regions of an RNA molecule (that is, not involved in formation of a double helix), it can chemically attack the adjacent phosphodiester bond to cleave the backbone.

RNA is transcribed with only four bases (adenine, cytosine, guanine and uracil), but these bases and attached sugars can be modified in numerous ways as the RNAs mature. Pseudouridine (Ø), in which the linkage between uracil and ribose is changed from a C–N bond to a

C–C bond, and ribothymidine (T) are found in various places (the most notable ones being in the TØC loop of tRNA). Another notable modified base is hypoxanthine, a deaminated adenine base whose nucleoside is called inosine (I). Inosine plays a key role in the wobble hypothesis of the genetic code.

There are nearly 100 other naturally occurring modified nucleosides, of which pseudouridine and nucleosides with 2'-O-methylribose are the most common. The specific roles of many of these modifications in RNA are not fully understood. However, it is notable that, in ribosomal RNA, many of the post-transcriptional modifications occur in highly functional regions, such as the peptidyl transferase centre and the subunit interface, implying that they are important for normal function.

The functional form of single stranded RNA molecules, just like proteins, frequently requires a specific tertiary structure. The scaffold for this structure is provided by secondary structural elements that are hydrogen bonds within the molecule. This leads to several recognizable "domains" of secondary structure like hairpin loops, bulges, and internal loops. Since RNA is charged, metal ions such as Mg2+ are needed to stabilise many secondary and tertiary structures.

Synthesis

Synthesis of RNA is usually catalyzed by an enzyme—RNA polymerase—using DNA as a template, a process known as transcription. Initiation of transcription begins with the binding of the enzyme to a promoter sequence in the DNA (usually found "upstream" of a gene). The DNA double helix is unwound by the helicase activity of the enzyme. The enzyme then progresses along the template strand in the 3' to 5' direction, synthesizing a complementary RNA molecule with elongation occurring in the 5' to 3' direction.

The DNA sequence also dictates where termination of RNA synthesis will occur. RNAs are often modified by enzymes after transcription. For example, a poly(A) tail and a 5' cap are added to eukaryotic pre-mRNA and introns are removed by the spliceosome. There are also a number of RNA-dependent RNA polymerases that use RNA as their template for synthesis of a new strand of RNA. For instance, a number of RNA viruses (such as poliovirus) use this type of enzyme to replicate their genetic material. Also, RNA-dependent RNA polymerase is part of the RNA interference pathway in many organisms.

Types of RNA

Messenger RNA (mRNA) is the RNA that carries information from DNA to the ribosome, the sites of protein synthesis (translation) in the cell. The coding sequence of the mRNA determines the amino acid sequence in the protein that is produced. Many RNAs do not code for protein however (about 97% of the transcriptional output is non-protein-coding in eukaryotes).

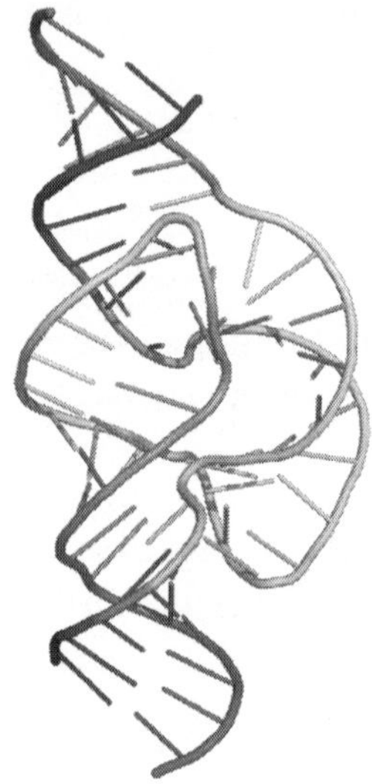

Figure 14: Structure of a hammerhead ribozyme, a ribozyme that cuts RNA

These so-called non-coding RNAs ("ncRNA") can be encoded by their own genes (RNA genes), but can also derive from mRNA introns. The most prominent examples of non-coding RNAs are transfer RNA (tRNA) and ribosomal RNA (rRNA), both of which are involved in the process of translation. There are also non-coding RNAs involved in gene regulation, RNA processing and other roles. Certain RNAs are able to catalyse chemical reactions such as cutting and ligating other RNA molecules, and the catalysis of peptide bond formation in the ribosome; these are known as ribozymes.

In Translation: Messenger RNA (mRNA) carries information about a protein sequence to the ribosomes, the protein synthesis factories in the cell. It is coded so that every three nucleotides (a codon) correspond to one amino acid. In eukaryotic cells, once precursor mRNA (pre-mRNA) has been transcribed from DNA, it is processed to mature mRNA. This removes its introns—non-coding sections of the pre-mRNA. The mRNA is then exported from the nucleus to the cytoplasm, where it is bound to ribosomes and translated into its corresponding protein form with the help of tRNA. In prokaryotic cells, which do not have nucleus and cytoplasm compartments, mRNA can bind to ribosomes while it is being transcribed from DNA. After a certain amount of time the message degrades into its component nucleotides with the assistance of ribonucleases.

Transfer RNA (tRNA) is a small RNA chain of about 80 nucleotides that transfers a specific amino acid to a growing polypeptide chain at the ribosomal site of protein synthesis during translation. It has sites for amino acid attachment and an anticodon region for codon recognition that binds to a specific sequence on the messenger RNA chain through hydrogen bonding.

Ribosomal RNA (rRNA) is the catalytic component of the ribosomes. Eukaryotic ribosomes contain four different rRNA molecules: 18S, 5.8S, 28S and 5S rRNA. Three of the rRNA molecules are synthesized in the nucleolus, and one is synthesized elsewhere. In the cytoplasm, ribosomal RNA and protein combine to form a nucleoprotein called a ribosome. The ribosome binds mRNA and carries out protein synthesis. Several ribosomes may be attached to a single mRNA at any time. rRNA is extremely abundant and makes up 80% of the 10 mg/ml RNA found in a typical eukaryotic cytoplasm.

Transfer-messenger RNA (tmRNA) is found in many bacteria and plastids. It tags proteins encoded by mRNAs that lack stop codons for degradation and prevents the ribosome from stalling.

Regulatory RNAs: Several types of RNA can downregulate gene expression by being complementary to a part of an mRNA or a gene's DNA. MicroRNAs (miRNA; 21-22 nt) are found in eukaryotes and act through RNA interference (RNAi), where an effector complex of miRNA and enzymes can break down mRNA to which the miRNA is complementary, block the mRNA from being translated, or accelerate its degradation. While small interfering RNAs (siRNA; 20-25 nt) are often produced by breakdown of viral RNA, there are also endogenous sources of siRNAs.

siRNAs act through RNA interference in a fashion similar to miRNAs. Some miRNAs and siRNAs can cause genes they target to be methylated, thereby decreasing or increasing transcription of those genes. Animals have Piwi-interacting RNAs (piRNA; 29-30 nt) which are active in germline cells and are thought to be a defense against transposons and play a role in gametogenesis.

Many prokaryotes have CRISPR RNAs, a regulatory system similar to RNA interference. Antisense RNAs are widespread; most downregulate a gene, but a few are activators of transcription. One way antisense RNA can act is by binding to an mRNA, forming double-stranded RNA that is enzymatically degraded. There are many long noncoding RNAs that regulate genes in eukaryotes, one such RNA is Xist, which coats one X chromosome in female mammals and inactivates it.

An mRNA may contain regulatory elements itself, such as riboswitches, in the 5' untranslated region or 3' untranslated region; these cis-regulatory elements regulate the activity of that mRNA. The untranslated regions can also contain elements that regulate other genes.

In RNA Processing: Many RNAs are involved in modifying other RNAs. Introns are spliced out of pre-mRNA by spliceosomes, which contain several small nuclear RNAs (snRNA), or the introns can be ribozymes that are spliced by themselves. RNA can also be altered by having its nucleotides modified to other nucleotides than A, C, G and U.

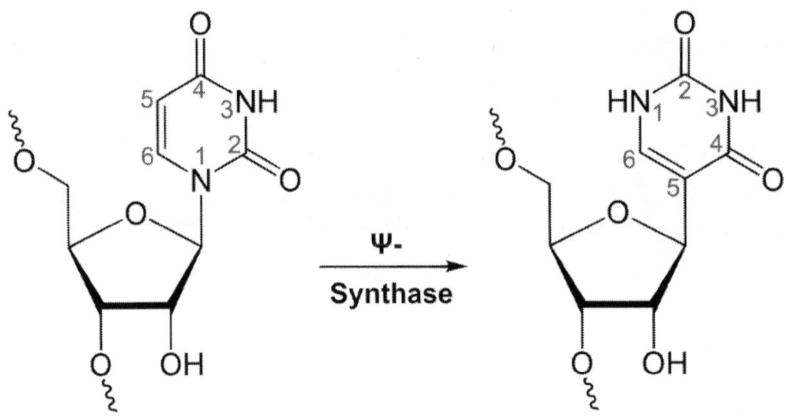

Figure 15: Uridine to pseudouridine is a common RNA modification.

In eukaryotes, modifications of RNA nucleotides are generally directed by small nucleolar RNAs (snoRNA; 60-300 nt), found in the nucleolus and cajal bodies. snoRNAs associate with enzymes and guide them to a spot on an RNA by basepairing to that RNA. These enzymes then perform the nucleotide modification. rRNAs and tRNAs are extensively modified, but snRNAs and mRNAs can also be the target of base modification.

RNA Genomes: Like DNA, RNA can carry genetic information. RNA viruses have genomes composed of RNA, and a variety of proteins encoded by that genome. The viral genome is replicated by some of those proteins, while other proteins protect the genome as the virus particle moves to a new host cell. Viroids are another group of pathogens, but they consist only of RNA, do not encode any protein and are replicated by a host plant cell's polymerase.

In Reverse Transcription: Reverse transcribing viruses replicate their genomes by reverse transcribing DNA copies from their RNA; these DNA copies are then transcribed to new RNA. Retrotransposons also spread by copying DNA and RNA from one another, and telomerase contains an RNA that is used as template for building the ends of eukaryotic chromosomes.

Double-Stranded RNA: Double-stranded RNA (dsRNA) is RNA with two complementary strands, similar to the DNA found in all cells. dsRNA forms the genetic material of some viruses (double-stranded RNA viruses). Double-stranded RNA such as viral RNA or siRNA can trigger RNA interference in eukaryotes, as well as interferon response in vertebrates.

Key Discoveries in RNA Biology

Research on RNA has led to many important biological discoveries and numerous Nobel Prizes. Nucleic acids were discovered in 1868 by Friedrich Miescher, who called the material 'nuclein' since it was found in the nucleus. It was later discovered that prokaryotic cells, which do not have a nucleus, also contain nucleic acids. The role of RNA in protein synthesis was suspected already in 1939. Severo Ochoa won the 1959 Nobel Prize in Medicine (shared with Arthur Kornberg) after he discovered an enzyme that can synthesize RNA in the laboratory. Ironically, the enzyme discovered by Ochoa (polynucleotide phosphorylase) was later shown to be responsible for RNA degradation, not RNA synthesis. The sequence of the 77 nucleotides of a yeast tRNA was found by Robert W. Holley in 1965, winning Holley the 1968 Nobel Prize in Medicine (shared with Har Gobind Khorana and Marshall Nirenberg). In 1967, Carl Woese hypothesized that RNA might be catalytic and suggested that the earliest forms of life (self-replicating molecules) could have relied on RNA both to carry genetic information and to catalyze biochemical reactions—an RNA world.

During the early 1970s, retroviruses and reverse transcriptase were discovered, showing for the first time that enzymes could copy RNA into DNA (the opposite of the usual route for transmission of genetic information). For this work, David Baltimore, Renato Dulbecco and Howard Temin were awarded a Nobel Prize in 1975. In 1976, Walter Fiers and his team determined the first complete nucleotide sequence of an RNA virus genome, that of bacteriophage MS2.

In 1977, introns and RNA splicing were discovered in both mammalian viruses and in cellular genes, resulting in a 1993 Nobel to Philip Sharp and Richard Roberts. Catalytic RNA molecules (ribozymes) were discovered in the early 1980s, leading to a 1989 Nobel award to Thomas Cech and Sidney Altman. In 1990 it was found in petunia that introduced genes can silence similar genes of the plant's own, now known to be a result of RNA interference. At about the same time, 22 nt long RNAs, now called microRNAs, were found to have a role in the development of *C. elegans*. Studies on RNA interference gleaned a Nobel Prize for Andrew Fire and Craig Mello in 2006, and another Nobel was awarded for studies on transcription of RNA to Roger Kornberg in the same year. The discovery of gene regulatory RNAs has led to attempts to develop drugs made of RNA, such as siRNA, to silence genes.

Nucleotide

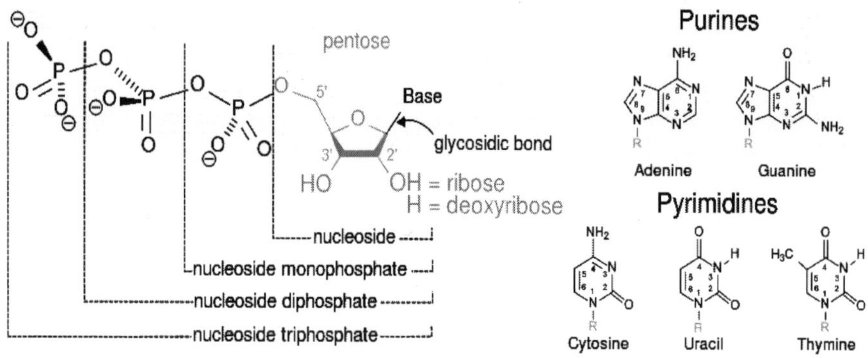

Figure 16: Structural elements of the most common nucleotides

Nucleotides are molecules that, when joined together, make up the structural units of RNA and DNA. In addition, nucleotides play central roles in metabolism. In that capacity, they serve as sources of chemical energy (adenosine triphosphate and guanosine triphosphate), participate in cellular signaling (cyclic guanosine monophosphate and cyclic adenosine monophosphate), and are incorporated into important cofactors of enzymatic reactions (coenzyme A, flavin adenine dinucleotide, flavin mononucleotide, and nicotinamide adenine dinucleotide phosphate).

Nucleotide Structure

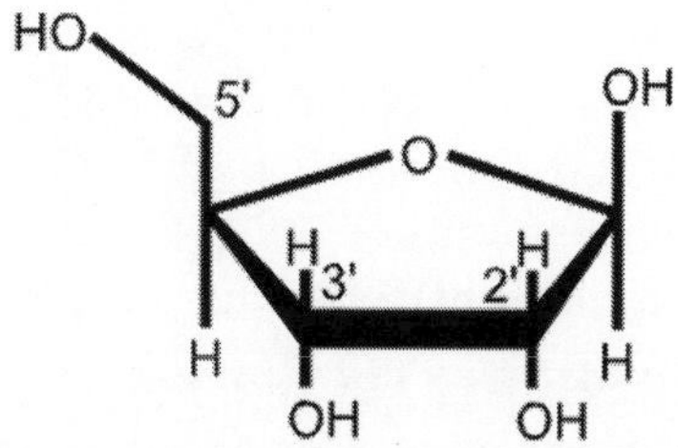

Figure 17: Ribose structure indicating numbering of carbon atoms

A nucleotide is composed of a nucleobase (nitrogenous base), a five-carbon sugar (either ribose or 2'-deoxyribose), and one to three phosphate groups. Together, the nucleobase and sugar comprise a nucleoside. The phosphate groups form bonds with either the 2, 3, or 5-carbon of the sugar, with the 5-carbon site most common. Cyclic nucleotides form when the phosphate group is bound to two of the sugar's hydroxyl groups. Ribonucleotides are nucleotides where the sugar is ribose, and deoxyribonucleotides contain the sugar deoxyribose. Nucleotides can contain either a purine or a pyrimidine base. Nucleic acids are polymeric macromolecules made from nucleotide monomers. In DNA, the purine bases are adenine and guanine, while the pyrimidines are thymine and cytosine. RNA uses uracil in place of thymine. Adenine always pairs with thymine by 2 hydrogen bonds, while guanine pairs with cytosine through 3 hydrogen bonds, each due to their unique structures.

Synthesis

Nucleotides can be synthesized by a variety of means both in vitro and in vivo. In vivo, nucleotides can be synthesised de novo or recycled through salvage pathways. Nucleotides undergo breakdown such that useful parts can be reused in synthesis reactions to create new nucleotides. In vitro, protecting groups may be used during laboratory production of nucleotides. A purified nucleoside is protected to create a phosphoramidite, which can then be used to obtain analogues not found in nature and/or to synthesize an oligonucleotide.

Pyrimidine Ribonucleotides: Pyrimidine nucleotide synthesis starts with the formation of carbamoyl phosphate from glutamine and CO_2. The cyclisation reaction between carbamoyl phosphate reacts with aspartate, yielding orotate in subsequent steps. Orotate reacts with 5-phosphoribosyl á-diphosphate (PRPP), yielding orotidine monophosphate (OMP), which is decarboxylated to form uridine monophosphate (UMP). It is from UMP that other pyrimidine nucleotides are derived. UMP is phosphorylated to uridine triphosphate (UTP) via two sequential reactions with ATP. Cytidine monophosphate (CMP) is derived from conversion of UTP to cytidine triphosphate (CTP) with subsequent loss of two phosphates.

Purine Ribonucleotides: The atoms which are used to build the purine nucleotides come from a variety of sources:

- The biosynthetic origins of purine ring atoms
- N1 arises from the amine group of Asp
- C2 and C8 originate from formate
- N3 and N9 are contributed by the amide group of Gln
- C4, C5 and N7 are derived from Gly
- C6 comes from HCO_3^- (CO_2)

The de novo synthesis of purine nucleotides by which these precursors are incorporated into the purine ring proceeds by a 10-step pathway to the branch-point intermediate IMP, the nucleotide of the base hypoxanthine. AMP and GMP are subsequently synthesized from this intermediate via separate, two-step pathways. Thus, purine moieties are initially formed as part of the ribonucleotides rather than as free bases.

Six enzymes take part in IMP synthesis. Three of them are multifunctional:

- GART (reactions 2, 3, and 5)
- PAICS (reactions 6, and 7)
- ATIC (reactions 9, and 10)

Reaction 1. The pathway starts with the formation of PRPP. PRPS1 is the enzyme that activates R5P, which is formed primarily by the pentose phosphate pathway, to PRPP by reacting it with ATP. The reaction is unusual in that a pyrophosphoryl group is directly transferred from ATP to C1 of R5P and that the product has the á configuration about C1. This reaction is also shared with the pathways for the synthesis of the pyrimidine nucleotides, Trp, and His. As a result of being on (a) such (a) major metabolic crossroad and the use of energy, this reaction is highly regulated.

Reaction 2. In the first reaction unique to purine nucleotide biosynthesis, PPAT catalyzes the displacement of PRPP's pyrophosphate group (PP_i) by Gln's amide nitrogen. The reaction occurs with the inversion of configuration about ribose C1, thereby forming â-5-phosphorybosylamine (5-PRA) and establishing the anomeric form of the future nucleotide. This reaction, which is driven to completion by the subsequent hydrolysis of the released PP_i, is the pathway's flux-generating step and is therefore regulated, too.

Length Unit

Nucleotide (abbreviated nt) is a common length unit for single-stranded RNA, similar to how base pair is a length unit for double-stranded DNA.

Gluconeogenesis

Gluconeogenesis (abbreviated GNG) is a metabolic pathway that results in the generation of glucose from non-carbohydrate carbon substrates such as lactate, glycerol, and glucogenic amino acids. It is one of the two main mechanisms humans and many other animals use to keep blood glucose levels from dropping too low (hypoglycemia). The other means of maintaining blood glucose levels is through the degradation of glycogen (glycogenolysis). Gluconeogenesis is a ubiquitous process, present in plants, animals, fungi, bacteria, and other microorganisms.

In animals, gluconeogenesis takes place mainly in the liver and, to a lesser extent, in the cortex of kidneys. This process occurs during periods of fasting, starvation, low-carbohydrate diets, or intense exercise and is highly endergonic. For example, the pathway leading from phosphoenolpyruvate to glucose-6-phosphate requires 6 molecules of ATP. Gluconeogenesis is often associated with ketosis. Gluconeogenesis is also a target of therapy for type II diabetes, such as metformin, which inhibits glucose formation and stimulates glucose uptake by cells.

Entering the Pathway

Lactate is transported back to the liver where it is converted into pyruvate by the Cori cycle using the enzyme lactate dehydrogenase. Pyruvate, the first designated substrate of the gluconeogenic pathway, can then be used to generate glucose. All citric acid cycle intermediates, through conversion to oxaloacetate, amino acids other than lysine or leucine, and glycerol can also function as substrates for gluconeogenesis. Transamination or deamination of amino acids facilitates entering of their carbon skeleton into the cycle directly (as pyruvate or oxaloacetate), or indirectly via the citric acid cycle.

Whether fatty acids can be converted into glucose in animals has been a longstanding question in biochemistry. It is known that odd-chain fatty acids can be oxidized to yield propionyl CoA, a precursor for succinyl CoA, which can be converted to pyruvate and enter into gluconeogenesis. In plants, to be specific, in seedlings, the glyoxylate cycle can be used to convert fatty acids (acetate) into the primary carbon source of the organism. The glyoxylate cycle produces four-carbon dicarboxylic acids that can enter gluconeogenesis.

In 1995, researchers identified the glyoxylate cycle in nematodes. In addition, the glyoxylate enzymes malate synthase and isocitrate lyase have been found in animal tissues. Genes coding for malate synthase gene have been identified in other [metazoans] including arthropods, echinoderms, and even some vertebrates. Mammals found to possess these genes include monotremes (platypus) and marsupials (opossum) but not placental mammals. Genes for isocitrate lyase are found only in nematodes, in which, it is apparent, they originated in horizontal gene transfer from bacteria.

The existence of glyoxylate cycles in humans has not been established, and it is widely held that fatty acids cannot be converted to glucose in humans directly. However, carbon-14 has been shown to end up in glucose when it is supplied in fatty acids. Despite these findings, it is considered unlikely that the 2-carbon acetyl-CoA derived from the oxidation of fatty acids would produce a net yield of glucose via the citric acid cycle. However, it is possible that, with additional sources of carbon via other pathways, glucose could be synthesized from acetyl-CoA. In fact, it is known that Ketone bodies, â-hydroxybutyrate in particular, can be converted to glucose at least in small amounts (â-hydroxybutyrate to acetoacetate to acetone to propanediol to pyruvate to glucose). Glycerol, which is a part of the triacylglycerol molecule, can be used in gluconeogenesis.

Location

In humans, gluconeogenesis is restricted to the liver and to a lesser extent the kidney. In all species, the formation of oxaloacetate from pyruvate and TCA cycle intermediates is restricted to the mitochondrion, and the enzymes that convert PEP to glucose are found in the cytosol. The location of the enzyme that links these two parts of gluconeogenesis by converting oxaloacetate to PEP, PEP carboxykinase, is variable by species: it can be found entirely within the mitochondria, entirely within the cytosol, or dispersed evenly between the two, as it is in humans.

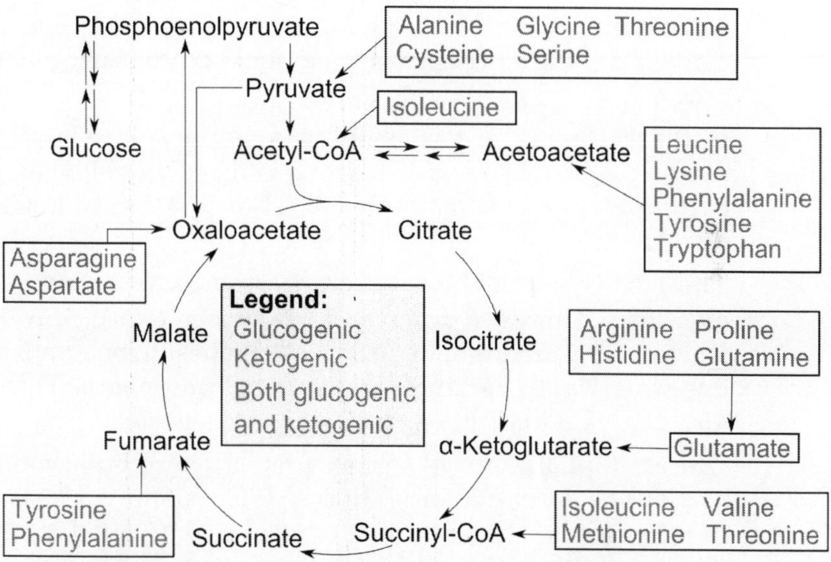

Figure 18: Catabolism of proteinogenic amino acids. Amino acids are classified according the abilities of their products to enter gluconeogenesis: Glucogenic amino acids have this ability; Ketogenic amino acids do not. These products may still be used for ketogenesis or lipid synthesis; Some amino acids are catabolized into both glucogenic and ketogenic products.

Transport of PEP across the mitochondrial membrane is accomplished by dedicated transport proteins; however no such proteins exist for oxaloacetate. Therefore species that lack intra-mitochondrial PEP, oxaloacetate must be converted into malate or asparate, exported from the mitochondrion, and converted back into oxaloacetate in order to allow gluconeogenesis to continue.

Pathway

Gluconeogenesis is a pathway consisting of eleven enzyme-catalyzed reactions. The pathway can begin in the mitochondria or cytoplasm, depending on the substrate being used. Many of the reactions are the reversible steps found in glycolysis.

- Gluconeogenesis begins in the mitochondria with the formation of oxaloacetate through carboxylation of pyruvate.

This reaction also requires one molecule of ATP, and is catalyzed by pyruvate carboxylase. This enzyme is stimulated by high levels of acetyl-CoA (produced in â-oxidation in the liver) and inhibited by high levels of ADP.

- Oxaloacetate is reduced to malate using NADH, a step required for transport out of the mitochondria.
- Malate is oxidized to oxaloacetate using NAD^+ in the cytoplasm, where the remaining steps of gluconeogenesis occur.
- Oxaloacetate is decarboxylated and phosphorylated to produce phosphoenolpyruvate by phosphoenolpyruvate carboxykinase. One molecule of GTP is hydrolyzed to GDP during this reaction.
- The next steps in the reaction are the same as reversed glycolysis. However, fructose-1,6-bisphosphatase converts fructose-1,6-bisphosphate to fructose 6-phosphate, requiring one water molecule and releasing one phosphate. This is also the rate-limiting step of gluconeogenesis.
- Glucose-6-phosphate is formed from fructose 6-phosphate by phosphoglucoisomerase. Glucose-6-phosphate can be used in other metabolic pathways or dephosphorylated to free glucose. Whereas free glucose can easily diffuse in and out of the cell, the phosphorylated form (glucose-6-phosphate) is locked in the cell, a mechanism by which intracellular glucose levels are controlled by cells.
- The final reaction of gluconeogenesis, the formation of glucose, occurs in the lumen of the endoplasmic reticulum, where glucose-6-phosphate is hydrolyzed by glucose-6-phosphatase to produce glucose. Glucose is shuttled into the cytosol by glucose transporters located in the membrane of the endoplasmic reticulum.

Regulation

While most steps in gluconeogenesis are the reverse of those found in glycolysis, three regulated and strongly exergonic reactions are replaced with more kinetically favourable reactions. Hexokinase/ glucokinase, phosphofructokinase, and pyruvate kinase enzymes of glycolysis are replaced with glucose-6-phosphatase, fructose-1,6-bisphosphatase, and PEP carboxykinase. This system of reciprocal control allow glycolysis and gluconeogenesis to inhibit each other and prevent the formation of a futile cycle. The majority of the enzymes responsible for gluconeogenesis are found in the cytoplasm; the exceptions are mitochondrial pyruvate carboxylase and, in animals, phosphoenolpyruvate carboxykinase. The latter exists as an isozyme located in both the mitochondrion and the cytosol. The rate of gluconeogenesis is ultimately controlled by the action of a key enzyme, fructose-1,6-bisphosphatase, which is also regulated through signal tranduction by cAMP and its phosphorylation.

Most factors that regulate the activity of the gluconeogenesis pathway do so by inhibiting the activity or expression of key enzymes. However, both acetyl CoA and citrate activate gluconeogenesis enzymes (pyruvate carboxylase and fructose-1,6-bisphosphatase, respectively). Due to the reciprocal control of the cycle, acetyl-CoA and citrate also have inhibitory roles in the activity of pyruvate kinase. Global control of gluconeogenesis is mediated by glucagon (*released when blood glucose is low*); it triggers phosphorylation of enzymes and regulatory proteins by Protein Kinase A (a cyclic AMP regulated kinase) resulting in inhibition of glycolysis and stimulation of gluconeogenesis, thus *bringing blood glucose levels up.*

Chapter 3

Understanding Genetics and Genetic Engineering

Immunology

Immunology is a broad branch of biomedical science that covers the study of all aspects of the immune system in all organisms. It deals with the physiological functioning of the immune system in states of both health and disease; malfunctions of the immune system in immunological disorders (autoimmune diseases, hypersensitivities, immune deficiency, transplant rejection); the physical, chemical and physiological characteristics of the components of the immune system in vitro, in situ, and in vivo. Immunology has applications in several disciplines of science, and as such is further divided.

Histological Examination of the Immune System

Even before the concept of immunity (from *immunis*, Latin for "exempt") was developed, numerous early physicians characterized organs that would later prove to be part of the immune system. The key primary lymphoid organs of the immune system are like thymus and bone marrow, and secondary lymphatic tissues such as spleen, tonsils, lymph vessels, lymph nodes, adenoids, and skin and liver. When health conditions warrant, immune system organs including the thymus, spleen, portions of bone marrow, lymph nodes and secondary lymphatic tissues can be surgically excised for examination while patients are still alive. Many components of the immune system are actually cellular in nature and not associated with any specific organ but rather are embedded or circulating in various tissues located throughout the body.

Classical Immunology

Classical immunology ties in with the fields of epidemiology and medicine. It studies the relationship between the body systems, pathogens, and immunity. The earliest written mention of immunity can be traced back to the plague of Athens in 430 BCE. Thucydides noted that people who had recovered from a previous bout of the disease could nurse the sick without contracting the illness a second time. Many other ancient societies have references to this phenomenon, but it was not until the 19th and 20th centuries before the concept developed into scientific theory.

The study of the molecular and cellular components that comprise the immune system, including their function and interaction, is the central science of immunology. The immune system has been divided into a more primitive innate immune system, and acquired or adaptive immune system of vertebrates, the latter of which is further divided into humoral and cellular components.

The humoral (antibody) response is defined as the interaction between antibodies and antigens. Antibodies are specific proteins released from a certain class of immune cells (B lymphocytes). Antigens are defined as anything that elicits generation of antibodies, hence they are Antibody Generators. Immunology itself rests on an understanding of the properties of these two biological entities. However, equally important is the cellular response, which can not only kill infected cells in its own right, but is also crucial in controlling the antibody response. Put simply, both systems are highly interdependent. In the 21st century, immunology has broadened its horizons with much research being performed in the more specialized niches of immunology. This includes the immunological function of cells, organs and systems not normally associated with the immune system, as well as the function of the immune system outside classical models of immunity (Yemeserach 2010).

Clinical Immunology

Clinical immunology is the study of diseases caused by disorders of the immune system (failure, aberrant action, and malignant growth of the cellular elements of the system). It also involves diseases of other systems, where immune reactions play a part in the pathology and clinical features.

The diseases caused by disorders of the immune system fall into two broad categories: immunodeficiency, in which parts of the immune system fail to provide an adequate response (examples include chronic granulomatous disease), and autoimmunity, in which the immune system attacks its own host's body (examples include systemic lupus erythematosus, rheumatoid arthritis, Hashimoto's disease and myasthenia gravis). Other immune system disorders include different hypersensitivities, in which the system responds inappropriately to harmless compounds (asthma and other allergies) or responds too intensely.The study of the molecular and cellular components that comprise the immune system, including their function and interaction, is the central science of immunology. The immune system has been divided into a more primitive.

The most well-known disease that affects the immune system itself is AIDS, caused by HIV. AIDS is an immunodeficiency characterized by the lack of CD4+ ("helper") T cells and macrophages, which are destroyed by HIV.

Clinical immunologists also study ways to prevent transplant rejection, in which the immune system attempts to destroy allografts os.

Developmental Immunology

The body's capability to react to antigen depends on a person's age, antigen type, maternal factors and the area where the antigen is presented. Neonates are said to be in a state of physiological immunodeficiency, because both their innate and adaptive immunological responses are greatly suppressed. Once born, a child's immune system responds favourably to protein antigens while not as well to glycoproteins and polysaccharies. In fact, many of the infections acquired by neonates are caused by low virulence organisms like Staphylococcus and Pseudomonas. In neonates, opsonic activity and the ability to activate the complement cascade is very limited. For example, the mean level of C3 in a newborn is approximately 65% of that found in the adult. Phagocytic activity is also greatly impaired in newborns. This is due to lower opsonic activity, as well as diminished up-regulation of integrin and selectin receptors, which limit the ability of neutrophils to interact with adhesion molecules in the endothelium. Their monocytes are slow and have a reduced ATP production, which also limits the newborns phagocitic activity. Although, the number of total lymphocytes is significantly higher than in adults, the cellular and humoral immunity is also impaired. Antigen presenting cells in newborns have a reduced capability to activate T cells. Also, T cells of a newborn proliferate poorly and produce very small amounts of cytokines like IL-2, IL-4, IL-5, IL-12, and IFN-g which limits their capacity to activate the humoral response as well as the phagocitic activity of macrophage. B cells develop early in gestation but are not fully active. Maternal factors also play a role in the body's immune response. At birth most of the immunoglobulin is present is maternal IgG. Because IgM, IgD, IgE and IgA don't cross the placenta, they are almost undetectable at birth. Although some IgA is provided in breast milk. These passively acquired antibodies can protect the newborn up to 18 months, but their response is usually short-lived and of low affinity. These antibodies can also produce a negative response. If a child is exposed to the antibody for a particular antigen before being exposed to the antigen itself then the child will produce a dampened response.

Passively acquired maternal antibodies can suppress the antibody response to active immunization. Similarly the response of T-cells to vaccination differs in children compared to adults, and vaccines that induce Th1 responses in adults do not readily elicit these same responses in neonates. By 6-9 months after birth, a child's immune system begins to respond more strongly to glycoproteins. Not until 12-24 months of age is there a marked improvement in the body's response to polysaccharides. This can be the reason for the specific time frames found in vaccination schedules.

During adolescence the human body undergoes several physical, physiological and immunological changes. These changes are started and mediated by different hormones. Depending on the sex either testosterone or 17-â-oestradiol, act on male and female bodies accordingly, start acting at ages of 12 and 10 years. There is evidence that these steroids act directly not only on the primary and secondary sexual characteristics, but also have an effect on the development and regulation of the immune system. There is an increased risk in developing autoimmunity for pubescent and post pubescent females and males. There is also some evidence that cell surface receptors on B cells and macrophages may detect sex hormones in the system.

The female sex hormone 17-â-oestradiol has been shown to regulate the level of immunological response. Similarly, some male androgens, like testosterone, seem to suppress the stress response to infection; but other androgens like DHEA have the opposite effect, as it increases the immune response instead of down playing it. As in females, the male sex hormones seem to have more control of the immune system during puberty and the time right after than in fully developed adults. Other than hormonal changes physical changes like the involution of the Thymus during puberty will also affect the immunological response of the subject or patient.

Immunotherapy

The use of immune system components to treat a disease or disorder is known as immunotherapy. Immunotherapy is most commonly used in the context of the treatment of cancers together with chemotherapy (drugs) and radiotherapy (radiation). However, immunotherapy is also often used in the immunosuppressed (such as HIV patients) and people suffering from other immune deficiencies or autoimmune diseases.

Diagnostic Immunology

The specificity of the bond between antibody and antigen has made it an excellent tool in the detection of substances in a variety of diagnostic techniques. Antibodies specific for a desired antigen can be conjugated with a radiolabel, fluorescent label, or colour-forming enzyme and are used as a "probe" to detect it. However, the similarity between some antigens can lead to false positives and other errors in such tests by antibodies cross-reacting with antigens that aren't exact matches.

Evolutionary Immunology

Study of the immune system in extant species is capable of giving us a key understanding of the evolution of species and the immune system. A development of complexity of the immune system can be seen from simple phagocytotic protection of single celled organisms, to circulating antimicrobial peptides in insects to lymphoid organs in vertebrates. However, it is important to recognize that every organism living today has an immune system that has evolved to be absolutely capable of protecting it from most forms of harm; those organisms that did not adapt their immune systems to external threats are no longer around to be observed. Insects and other arthropods, while not possessing true adaptive immunity, show highly evolved systems of innate immunity, and are additionally protected from external injury (and exposure to pathogens) by their chitinous shells.

Reproductive Immunology

This area of the immunology is devoted to the study of immunological aspects of the reproductive process including fetus acceptance. The term has also been used by fertility clinics to address fertility problems, recurrent miscarriages, premature deliveries, and dangerous complications such as pre-eclampsia.

Immunologist

According to the American Academy of Allergy, Asthma, and Immunology (AAAAI), "an immunologist is a research scientist who investigates the immune system of vertebrates (including the human immune system). Immunologists include research scientists (Ph.D.) who work in laboratories. Immunologists also include physicians who, for example, treat patients with immune system disorders. Some immunologists are physician-scientists who combine laboratory research with patient care."

Genetics

Genetics, a discipline of biology, is the science of genes, heredity, and variation in living organisms. Genetics deals with the molecular structure and function of genes, with gene behaviour in the context of a cell or organism (e.g. dominance and epigenetics), with patterns of inheritance from parent to offspring, and with gene distribution, variation and change in populations. Given that genes are universal to living organisms, genetics can be applied to the study of all living systems, from viruses and bacteria, through plants (especially crops), to humans, as in medical genetics.

The fact that living things inherit traits from their parents has been used since prehistoric times to improve crop plants and animals through selective breeding. However, the modern science of genetics, which seeks to understand the process of inheritance, only began with the work of Gregor Mendel in the mid-19th century. Although he did not know the physical basis for heredity, Mendel observed that organisms inherit traits via discrete units of inheritance, which are now called genes.

Genes correspond to regions within DNA, a molecule composed of a chain of four different types of nucleotides—the sequence of these nucleotides is the genetic information organisms inherit. DNA naturally occurs in a double stranded form, with nucleotides on each strand complementary to each other. Each strand can act as a template for creating a new partner strand. This is the physical method for making copies of genes that can be inherited.

The sequence of nucleotides in a gene is translated by cells to produce a chain of amino acids, creating proteins—the order of amino acids in a protein corresponds to the order of nucleotides in the gene. This relationship between nucleotide sequence and amino acid sequence is known as the genetic code. The amino acids in a protein determine how it folds into a three-dimensional shape; this structure is, in turn, responsible for the protein's function. Proteins carry out almost all the functions needed for cells to live. A change to the DNA in a gene can change a protein's amino acds, changing its shape and function: this can have a dramatic effect in the cell and on the organism as a whole.

Although genetics plays a large role in the appearance and behaviour of organisms, it is the combination of genetics with what an organism experiences that determines the ultimate outcome. For example, while genes play a role in determining an organism's size, the nutrition and health it experiences after inception also have a large effect.

History

Although the science of genetics began with the applied and theoretical work of Gregor Mendel in the mid-19th century, other theories of inheritance preceded Mendel. A popular theory during Mendel's time was the concept of blending inheritance: the idea that individuals inherit a smooth blend of traits from their parents. Mendel's work disproved this, showing that traits are composed of combinations of distinct genes rather than a continuous blend. Another theory that had some support at that time was the inheritance of acquired characteristics: the belief that individuals inherit traits strengthened by their parents. This theory (commonly associated with Jean-Baptiste Lamarck) is now known to be wrong—the experiences of individuals do not affect the genes they pass to their children. Other theories included the pangenesis of Charles Darwin (which had both acquired and inherited aspects) and Francis Galton's reformulation of pangenesis as both particulate and inherited.

Mendelian and Classical Genetics: Modern genetics started with Gregor Johann Mendel, a German-Czech Augustinian monk and scientist who studied the nature of inheritance in plants. In his paper "Versuche über Pflanzenhybriden" ("Experiments on Plant Hybridization"), presented in 1865 to the *Naturforschender Verein* (Society for Research in Nature) in Brünn, Mendel traced the inheritance patterns of certain traits in pea plants and described them mathematically. Although this pattern of inheritance could only be observed for a few traits, Mendel's work suggested that heredity was particulate, not acquired, and that the inheritance patterns of many traits could be explained through simple rules and ratios.

The importance of Mendel's work did not gain wide understanding until the 1890s, after his death, when other scientists working on similar problems re-discovered his research. William Bateson, a proponent of Mendel's work, coined the word *genetics* in 1905. (The adjective *genetic*, derived from the Greek word *genesis*—ãÝíåóéò, "origin", predates the noun and was first used in a biological sense in 1860.) Bateson popularized the usage of the word *genetics* to describe the study of inheritance in his inaugural address to the Third International Conference on Plant Hybridization in London, England, in 1906.

After the rediscovery of Mendel's work, scientists tried to determine which molecules in the cell were responsible for inheritance. In 1910, Thomas Hunt Morgan argued that genes are on chromosomes, based on observations of a sex-linked white eye mutation in fruit flies. In 1913, his student Alfred Sturtevant used the phenomenon of genetic linkage to show that genes are arranged linearly on the chromosome.

Molecular Genetics

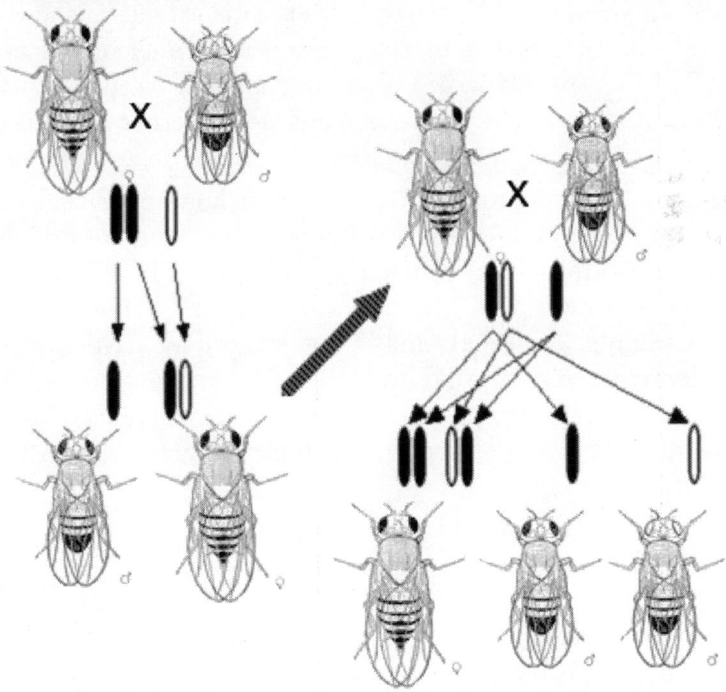

Figure 1: Morgan's observation of sex-linked inheritance of a mutation causing white eyes in Drosophila led him to the hypothesis that genes are located upon chromosomes.

Although genes were known to exist on chromosomes, chromosomes are composed of both protein and DNA—scientists did not know which of these is responsible for inheritance. In 1928, Frederick Griffith discovered the phenomenon of transformation: dead bacteria could transfer genetic material to "transform" other still-living bacteria. Sixteen years later, in 1944, Oswald Theodore Avery, Colin McLeod and Maclyn McCarty identified the molecule responsible for transformation as DNA. The Hershey-Chase experiment in 1952 also showed that DNA (rather than protein) is the genetic material of the viruses that infect bacteria, providing further evidence that DNA is the molecule responsible for inheritance.

James D. Watson and Francis Crick determined the structure of DNA in 1953, using the X-ray crystallography work of Rosalind Franklin and Maurice Wilkins that indicated DNA had a helical structure (i.e., shaped like a corkscrew). Their double-helix model had two strands of DNA with the nucleotides pointing inward, each matching a complementary nucleotide on the other strand to form what looks like rungs on a twisted ladder. This structure showed that genetic information exists in the sequence of nucleotides on each strand of DNA. The structure also suggested a simple method for duplication: if the strands are separated, new partner strands can be reconstructed for each based on the sequence of the old strand.

Although the structure of DNA showed how inheritance works, it was still not known how DNA influences the behaviour of cells. In the following years, scientists tried to understand how DNA controls the process of protein production. It was discovered that the cell uses DNA as a template to create matching messenger RNA (a molecule with nucleotides, very similar to DNA). The nucleotide sequence of a messenger RNA is used to create an amino acid sequence in protein; this translation between nucleotide and amino acid sequences is known as the genetic code.

With this molecular understanding of inheritance, an explosion of research became possible. One important development was chain-termination DNA sequencing in 1977 by Frederick Sanger. This technology allows scientists to read the nucleotide sequence of a DNA molecule. In 1983, Kary Banks Mullis developed the polymerase chain reaction, providing a quick way to isolate and amplify a specific section of a DNA from a mixture. Through the pooled efforts of the Human Genome Project and the parallel private effort by Celera

Genomics, these and other techniques culminated in the sequencing of the human genome in 2003.

Features of Inheritance

Discrete Inheritance and Mendel's Laws: At its most fundamental level, inheritance in organisms occurs by means of discrete traits, called genes. This property was first observed by Gregor Mendel, who studied the segregation of heritable traits in pea plants. In his experiments studying the trait for flower colour, Mendel observed that the flowers of each pea plant were either purple or white—but never an intermediate between the two colours. These different, discrete versions of the same gene are called alleles.

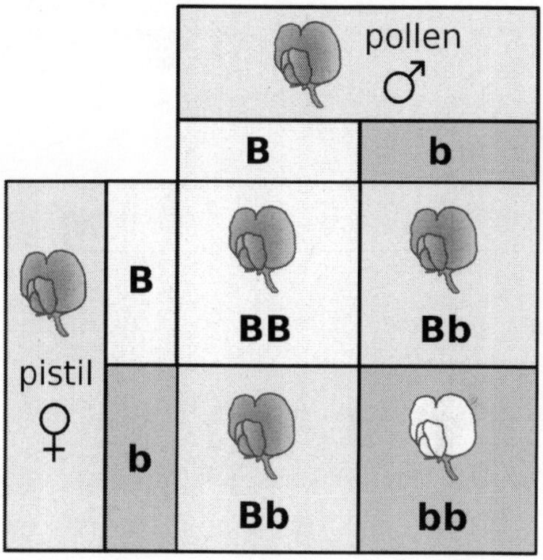

Figure 2: A Punnett square depicting a cross between two pea plants heterozygous for purple (B) and white (b) blossoms

In the case of pea, which is a diploid species, each individual plant has two alleles of each gene, one allele inherited from each parent. Many species, including humans, have this pattern of inheritance. Diploid organisms with two copies of the same allele of a given gene are called homozygous at that gene locus, while organisms with two different alleles of a given gene are called heterozygous.

The set of alleles for a given organism is called its genotype, while the observable traits of the organism are called its phenotype. When organisms are heterozygous at a gene, often one allele is called dominant as its qualities dominate the phenotype of the organism, while the other allele is called recessive as its qualities recede and are not observed. Some alleles do not have complete dominance and instead have incomplete dominance by expressing an intermediate phenotype, or codominance by expressing both alleles at once.

When a pair of organisms reproduce sexually, their offspring randomly inherit one of the two alleles from each parent. These observations of discrete inheritance and the segregation of alleles are collectively known as Mendel's first law or the Law of Segregation.

Notation and Diagrams

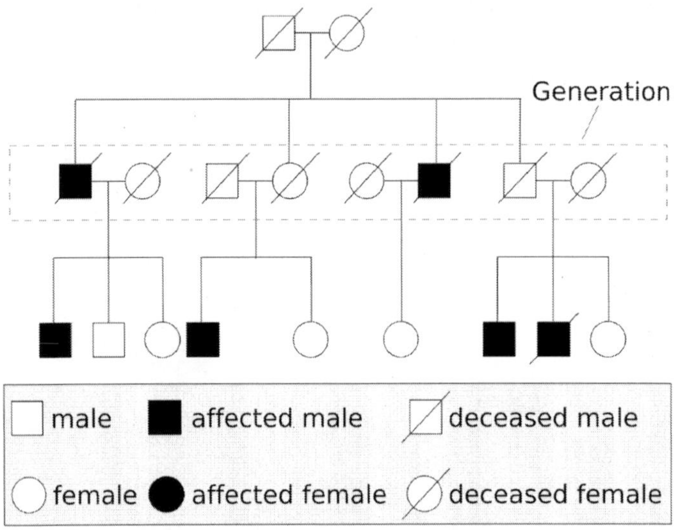

Figure 3: Genetic pedigree charts help track the inheritance patterns of traits.

Geneticists use diagrams and symbols to describe inheritance. A gene is represented by one or a few letters. Often a "+" symbol is used to mark the usual, non-mutant allele for a gene. In fertilization and breeding experiments (and especially when discussing Mendel's laws) the parents are referred to as the "P" generation and the offspring as the "F1" (first filial) generation. When the F1 offspring mate with each other, the offspring are called the "F2" (second filial) generation. One of the common diagrams used to predict the result of cross-breeding is the Punnett square. When studying human genetic diseases, geneticists often use pedigree charts to represent the inheritance of traits. These charts map the inheritance of a trait in a family tree.

Interactions of Multiple Genes

Organisms have thousands of genes, and in sexually reproducing organisms these genes generally assort independently of each other. This means that the inheritance of an allele for yellow or green pea colour is unrelated to the inheritance of alleles for white or purple flowers. This phenomenon, known as "Mendel's second law" or the "Law of independent assortment", means that the alleles of different genes get shuffled between parents to form offspring with many different combinations. (Some genes do not assort independently, demonstrating genetic linkage, a topic discussed later in this article.)

Often different genes can interact in a way that influences the same trait. In the Blue-eyed Mary (*Omphalodes verna*), for example, there exists a gene with alleles that determine the colour of flowers: blue or magenta. Another gene, however, controls whether the flowers have colour at all or are white. When a plant has two copies of this white allele, its flowers are white—regardless of whether the first gene has blue or magenta alleles. This interaction between genes is called epistasis, with the second gene epistatic to the first.

Many traits are not discrete features (e.g. purple or white flowers) but are instead continuous features (e.g. human height and skin colour). These complex traits are products of many genes. The influence of these genes is mediated, to varying degrees, by the environment an organism has experienced. The degree to which an organism's genes contribute to a complex trait is called heritability. Measurement of the heritability of a trait is relative—in a more variable environment, the environment has a bigger influence on the total variation of the trait. For example, human height is a trait with complex causes. It has a heritability of 89% in the United States. In Nigeria, however, where people experience a more variable access to good nutrition and health care, height has a heritability of only 62%.

Molecular Basis for Inheritance

DNA and Chromosomes: The molecular basis for genes is deoxyribonucleic acid (DNA). DNA is composed of a chain of nucleotides, of which there are four types: adenine (A), cytosine (C), guanine (G), and thymine (T). Genetic information exists in the sequence of these nucleotides, and genes exist as stretches of sequence along the DNA chain. Viruses are the only exception to this rule—sometimes viruses use the very similar molecule RNA instead of DNA as their genetic material.

DNA normally exists as a double-stranded molecule, coiled into the shape of a double-helix. Each nucleotide in DNA preferentially pairs with its partner nucleotide on the opposite strand: A pairs with T, and C pairs with G. Thus, in its two-stranded form, each strand effectively contains all necessary information, redundant with its partner strand. This structure of DNA is the physical basis for inheritance: DNA replication duplicates the genetic information by splitting the strands and using each strand as a template for synthesis of a new partner strand.

Genes are arranged linearly along long chains of DNA sequence, called chromosomes. In bacteria, each cell usually contains a single circular chromosome, while eukaryotic organisms (including plants and animals) have their DNA arranged in multiple linear chromosomes. These DNA strands are often extremely long; the largest human chromosome, for example, is about 247 million base pairs in length. The DNA of a chromosome is associated with structural proteins that organize, compact, and control access to the DNA, forming a material called chromatin; in eukaryotes, chromatin is usually composed of nucleosomes, segments of DNA wound around cores of histone proteins. The full set of hereditary material in an organism (usually the combined DNA sequences of all chromosomes) is called the genome.

While haploid organisms have only one copy of each chromosome, most animals and many plants are diploid, containing two of each chromosome and thus two copies of every gene. The two alleles for a gene are located on identical loci of sister chromatids, each allele inherited from a different parent. Many species have so called sex chromosomes. They are special in that they determine the sex of the organism. In humans and many other animals, the Y-chromosome contains the gene that triggers the development of the specifically male characteristics. In evolution, this chromosome has lost most of its content and also most of its genes, while the X chromosome is similar to the other chromosomes and contains many genes. The X and Y chromosomes form a very heterogeneous pair before cell division.

Reproduction: When cells divide, their full genome is copied and each daughter cell inherits one copy. This process, called mitosis, is the simplest form of reproduction and is the basis for asexual reproduction. Asexual reproduction can also occur in multicellular organisms, producing offspring that inherit their genome from a single parent. Offspring that are genetically identical to their parents are called clones. Eukaryotic organisms often use sexual reproduction to generate offspring that contain a mixture of genetic material inherited from two different parents. The process of sexual reproduction alternates between forms that contain single copies of the genome (haploid) and double copies (diploid). Haploid cells fuse and combine genetic material to create a diploid cell with paired chromosomes. Diploid organisms form haploids by dividing, without replicating their DNA, to create daughter cells that randomly inherit one of each pair of chromosomes. Most animals and many plants are diploid for most of their lifespan, with the haploid form reduced to single cell gametes such as sperm or eggs.

Although they do not use the haploid/diploid method of sexual reproduction, bacteria have many methods of acquiring new genetic information. Some bacteria can undergo conjugation, transferring a small circular piece of DNA to another bacterium. Bacteria can also take up raw DNA fragments found in the environment and integrate them into their genomes, a phenomenon known as transformation. These processes result in horizontal gene transfer, transmitting fragments of genetic information between organisms that would be otherwise unrelated.

Recombination and Linkage

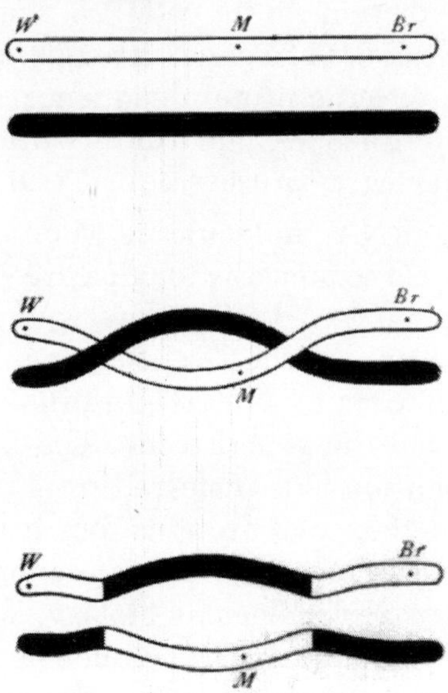

Figure 4: Thomas Hunt Morgan's 1916 illustration of a double crossover between chromosomes

The diploid nature of chromosomes allows for genes on different chromosomes to assort independently during sexual reproduction, recombining to form new combinations of genes. Genes on the same chromosome would theoretically never recombine, however, were it not for the process of chromosomal crossover. During crossover, chromosomes exchange stretches of DNA, effectively shuffling the gene alleles between the chromosomes. This process of chromosomal crossover generally occurs during meiosis, a series of cell divisions that creates haploid cells. The probability of chromosomal crossover occurring between two given points on the chromosome is related to the distance between the points. For an arbitrarily long distance, the probability of crossover is high enough that the inheritance of the genes is effectively uncorrelated. For genes that are closer together, however, the lower probability of crossover means that the genes demonstrate genetic linkage—alleles for the two genes tend to be inherited together. The amounts of linkage between a series of genes can be combined to form a linear linkage map that roughly describes the arrangement of the genes along the chromosome.

Gene Expression

Genetic Code: Genes generally express their functional effect through the production of proteins, which are complex molecules responsible for most functions in the cell. Proteins are chains of amino acids, and the DNA sequence of a gene (through an RNA intermediate) is used to produce a specific protein sequence. This process begins with the production of an RNA molecule with a sequence matching the gene's DNA sequence, a process called transcription.

This messenger RNA molecule is then used to produce a corresponding amino acid sequence through a process called translation. Each group of three nucleotides in the sequence, called a codon, corresponds either to one of the twenty possible amino acids in a protein or an instruction to end the amino acic sequence; this correspondence is called the genetic code. The flow of information is unidirectional: information is transferred from nucleotide sequences into the amino acid sequence of proteins, but it never transfers from protein back into the sequence of DNA—a phenomenon Francis Crick called the central dogma of molecular biology. The specific sequence of amino acids results in a unique three-dimensional structure for that protein, and the three-dimensional structures of proteins are related to their functions. Some are simple structural molecules, like the fibres formed by the protein collagen.

Proteins can bind to other proteins and simple molecules, sometimes acting as enzymes by facilitating chemical reactions within the bound molecules (without changing the structure of the protein itself). Protein structure is dynamic; the protein hemoglobin bends into slightly different forms as it facilitates the capture, transport, and release of oxygen molecules within mammalian blood. A single nucleotide difference within DNA can cause a change in the amino acid sequence of a protein. Because protein structures are the result of their amino acid sequences, some changes can dramatically change the properties of a protein by destabilizing the structure or changing the surface of the protein in a way that changes its interaction with other proteins and molecules. For example, sickle-cell anemia is a human genetic disease that results from a single base difference within the coding region for the â-globin section of hemoglobin, causing a single amino acid change that changes hemoglobin's physical properties. Sickle-cell versions of hemoglobin stick to themselves, stacking to form fibres that distort the shape of red blood cells carrying the protein. These sickle-shaped cells no longer flow smoothly through blood vessels, having a tendency to clog or degrade, causing the medical problems associated with this disease.

Some genes are transcribed into RNA but are not translated into protein products—such RNA molecules are called non-coding RNA. In some cases, these products fold into structures which are involved in critical cell functions (e.g. ribosomal RNA and transfer RNA). RNA can also have regulatory effect through hybridization interactions with other RNA molecules (e.g. microRNA).

Nature Versus Nurture: Although genes contain all the information an organism uses to function, the environment plays an important role in determining the ultimate phenotype—a phenomenon often referred to as "nature vs. nurture". The phenotype of an organism depends on the interaction of genetics with the environment. One example of this is the case of temperature-sensitive mutations. Often, a single amino acid change within the sequence of a protein does not change its behaviour and interactions with other molecules, but it does destabilize the structure. In a high temperature environment, where molecules are moving more quickly and hitting each other, this results in the protein losing its structure and failing to function. In a low temperature environment, however, the protein's structure is stable and it functions normally. This type of mutation is visible in the coat colouration of Siamese cats, where a mutation in an enzyme responsible for pigment production causes it to destabilize and lose function at high temperatures. The protein remains functional in areas of skin that are colder—legs, ears, tail, and face—and so the cat has dark fur at its extremities.

Environment also plays a dramatic role in effects of the human genetic disease phenylketonuria. The mutation that causes phenylketonuria disrupts the ability of the body to break down the amino acid phenylalanine, causing a toxic build-up of an intermediate molecule that, in turn, causes severe symptoms of progressive mental retardation and seizures. If someone with the phenylketonuria mutation follows a strict diet that avoids this amino acid, however, they remain normal and healthy.

A popular method to determine how much role nature and nurture play is to study identical and fraternal twins or siblings of multiple birth. Because identical siblings come from the same zygote they are genetically the same. Fraternal siblings however are as different genetically from one another as normal siblings. By comparing how often the twin of a set has the same disorder between fraternal and identical twins, scientists can see whether there is more of a nature or nurture effect. One famous example of a multiple birth study includes the Genain quadruplets, who were identical quadruplets all diagnosed with schizophrenia.

Gene Regulation: The genome of a given organism contains thousands of genes, but not all these genes need to be active at any given moment. A gene is expressed when it is being transcribed into mRNA (and translated into protein), and there exist many cellular methods of controlling the expression of genes such that proteins are produced only when needed by the cell.

Figure 5: Transcription factors bind to DNA, influencing the transcription of associated genes.

Transcription factors are regulatory proteins that bind to the start of genes, either promoting or inhibiting the transcription of the gene. Within the genome of *Escherichia coli* bacteria, for example, there exists a series of genes necessary for the synthesis of the amino acid tryptophan. However, when tryptophan is already available to the cell, these genes for tryptophan synthesis are no longer needed. The presence of tryptophan directly affects the activity of the genes—tryptophan molecules bind to the tryptophan repressor (a transcription factor), changing the repressor's structure such that the repressor binds to the genes. The tryptophan repressor blocks the transcription and expression of the genes, thereby creating negative feedback regulation of the tryptophan synthesis process.

Differences in gene expression are especially clear within multicellular organisms, where cells all contain the same genome but have very different structures and behaviours due to the expression of different sets of genes. All the cells in a multicellular organism derive from a single cell, differentiating into variant cell types in response to external and intercellular signals and gradually establishing different patterns of gene expression to create different behaviours. As no single gene is responsible for the development of structures within multicellular organisms, these patterns arise from the complex interactions between many cells.

Within eukaryotes there exist structural features of chromatin that influence the transcription of genes, often in the form of modifications to DNA and chromatin that are stably inherited by daughter cells. These features are called "epigenetic" because they exist "on top" of the DNA sequence and retain inheritance from one cell generation to the next. Because of epigenetic features, different cell types grown within the same medium can retain very different properties. Although epigenetic features are generally dynamic over the course of development, some, like the phenomenon of paramutation, have multigenerational inheritance and exist as rare exceptions to the general rule of DNA as the basis for inheritance.

Genetic Change

Mutations: During the process of DNA replication, errors occasionally occur in the polymerization of the second strand. These errors, called mutations, can have an impact on the phenotype of an organism, especially if they occur within the protein coding sequence of a gene. Error rates are usually very low—1 error in every 10–100 million bases—due to the "proofreading" ability of DNA polymerases. (Without proofreading error rates are a thousandfold higher; because many viruses rely on DNA and RNA polymerases that lack proofreading ability, they experience higher mutation rates.) Processes that increase the rate of changes in DNA are called mutagenic: mutagenic chemicals promote errors in DNA replication, often by interfering with the structure of base-pairing, while UV radiation induces mutations by causing damage to the DNA structure. Chemical damage to DNA occurs naturally as well, and cells use DNA repair mechanisms to repair mismatches and breaks in DNA—nevertheless, the repair sometimes fails to return the DNA to its original sequence.

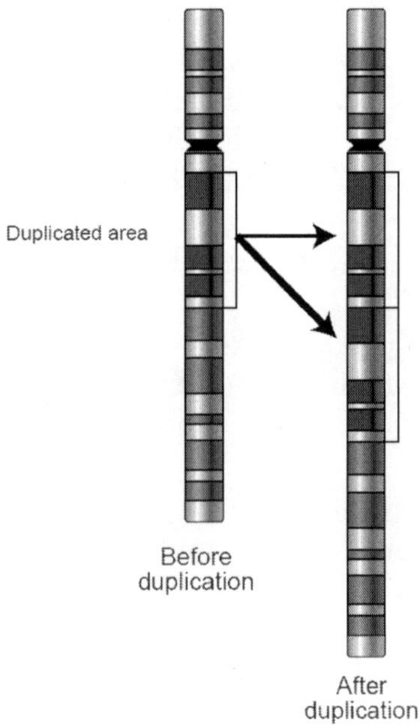

Figure 6: Gene duplication allows diversification by providing redundancy: one gene can mutate and lose its original function without harming the organism.

In organisms that use chromosomal crossover to exchange DNA and recombine genes, errors in alignment during meiosis can also cause mutations. Errors in crossover are especially likely when similar sequences cause partner chromosomes to adopt a mistaken alignment; this makes some regions in genomes more prone to mutating in this way. These errors create large structural changes in DNA sequence—duplications, inversions or deletions of entire regions, or the accidental exchanging of whole parts between different chromosomes (called translocation).

Natural Selection and Evolution: Mutations alter an organisms genotype and occasionally this causes different phenotypes to appear. Most mutations have little effect on an organism's phenotype, health, or reproductive fitness. Mutations that do have an effect are usually deleterious, but occasionally some can be beneficial. Studies in the fly *Drosophila melanogaster* suggest that if a mutation changes a protein produced by a gene, about 70 percent of these mutations will be harmful with the remainder being either neutral or weakly beneficial.

Population genetics studies the distribution of genetic differences within populations and how these distributions change over time. Changes in the frequency of an allele in a population are mainly influenced by natural selection, where a given allele provides a selective or reproductive advantage to the organism, as well as other factors such as genetic drift, artificial selection and migration. Over many generations, the genomes of organisms can change significantly, resulting in the phenomenon of evolution. Selection for beneficial mutations can cause a species to evolve into forms better able to survive in their environment, a process called adaptation. New species are formed through the process of speciation, often caused by geographical separations that prevent populations from exchanging genes with each other. The application of genetic principles to the study of population biology and evolution is referred to as the modern synthesis.

By comparing the homology between different species' genomes it is possible to calculate the evolutionary distance between them and when they may have diverged (called a molecular clock). Genetic comparisons are generally considered a more accurate method of characterizing the relatedness between species than the comparison of phenotypic characteristics. The evolutionary distances between species can be used to form evolutionary trees; these trees represent the common descent and divergence of species over time, although they do not show the transfer of genetic material between unrelated species (known as horizontal gene transfer and most common in bacteria).

Research and Technology

Model Organisms: Although geneticists originally studied inheritance in a wide range of organisms, researchers began to specialize in studying the genetics of a particular subset of organisms. The fact that significant research already existed for a given organism would encourage new researchers to choose it for further study, and so eventually a few model organisms became the basis for most genetics research. Common research topics in model organism genetics include the study of gene regulation and the involvement of genes in development and cancer.

Organisms were chosen, in part, for convenience—short generation times and easy genetic manipulation made some organisms popular genetics research tools. Widely used model organisms include the gut bacterium *Escherichia coli*, the plant *Arabidopsis thaliana*, baker's yeast (*Saccharomyces cerevisiae*), the nematode *Caenorhabditis elegans*, the common fruit fly (*Drosophila melanogaster*), and the common house mouse (*Mus musculus*).

Medicine: Medical genetics seeks to understand how genetic variation relates to human health and disease. When searching for an unknown gene that may be involved in a disease, researchers commonly use genetic linkage and genetic pedigree charts to find the location on the genome associated with the disease. At the population level, researchers take advantage of Mendelian randomization to look for locations in the genome that are associated with diseases, a technique especially useful for multigenic traits not clearly defined by a single gene. Once a candidate gene is found, further research is often done on the corresponding gene (called an orthologous gene) in model organisms. In addition to studying genetic diseases, the increased availability of genotyping techniques has led to the field of pharmacogenetics—studying how genotype can affect drug responses.

Individuals differ in their inherited tendency to develop cancer, and cancer is a genetic disease. The process of cancer development in the body is a combination of events. Mutations occasionally occur within cells in the body as they divide. Although these mutations will not be inherited by any offspring, they can affect the behaviour of cells, sometimes causing them to grow and divide more frequently. There are biological mechanisms that attempt to stop this process; signals are given to inappropriately dividing cells that should trigger cell death, but sometimes additional mutations occur that cause cells to ignore these messages. An internal process of natural selection occurs within the body and eventually mutations accumulate within cells to promote their own growth, creating a cancerous tumor that grows and invades various tissues of the body.

Research Techniques: DNA can be manipulated in the laboratory. Restriction enzymes are commonly used enzymes that cut DNA at specific sequences, producing predictable fragments of DNA.

DNA fragments can be visualized through use of gel electrophoresis, which separates fragments according to their length. The use of ligation enzymes allows DNA fragments to be connected, and by ligating fragments of DNA together from different sources, researchers can create recombinant DNA.

Often associated with genetically modified organisms, recombinant DNA is commonly used in the context of plasmids—short circular DNA fragments with a few genes on them. By inserting plasmids into bacteria and growing those bacteria on plates of agar (to isolate clones of bacteria cells), researchers can clonally amplify the inserted fragment of DNA (a process known as molecular cloning). (Cloning can also refer to the creation of clonal organisms, through various techniques.) DNA can also be amplified using a procedure called the polymerase chain reaction (PCR). By using specific short sequences of DNA, PCR can isolate and exponentially amplify a targeted region of DNA. Because it can amplify from extremely small amounts of DNA, PCR is also often used to detect the presence of specific DNA sequences.

DNA Sequencing and Genomics

One of the most fundamental technologies developed to study genetics, DNA sequencing allows researchers to determine the sequence of nucleotides in DNA fragments. Developed in 1977 by Frederick Sanger and coworkers, chain-termination sequencing is now routinely used to sequence DNA fragments. With this technology, researchers have been able to study the molecular sequences associated with many human diseases.

As sequencing has become less expensive, researchers have sequenced the genomes of many organisms, using computational tools to stitch together the sequences of many different fragments (a process called genome assembly). These technologies were used to sequence the human genome, leading to the completion of the Human Genome Project in 2003. New high-throughput sequencing technologies are dramatically lowering the cost of DNA sequencing, with many researchers hoping to bring the cost of resequencing a human genome down to a thousand dollars.

The large amount of sequence data available has created the field of genomics, research that uses computational tools to search for and analyze patterns in the full genomes of organisms. Genomics can also be considered a subfield of bioinformatics, which uses computational approaches to analyze large sets of biological data.

Genetic Engineering

Genetic engineering, also called genetic modification, is the direct human manipulation of an organism's genetic material in a way that does not occur under natural conditions. It involves the use of recombinant DNA techniques, but does not include traditional animal and plant breeding or mutagenesis. Any organism that is generated using these techniques is considered to be a genetically modified organism. The first organisms genetically engineered were bacteria in 1973 and then mice in 1974. Insulin producing bacteria were commercialized in 1982 and genetically modified food has been sold since 1994.

The most common form of genetic engineering involves the insertion of new genetic material at an unspecified location in the host genome. This is accomplished by isolating and copying the genetic material of interest, generating a construct containing all the genetic elements for correct expression, and then inserting this construct into the host organism. Other forms of genetic engineering include gene targeting and knocking out specific genes via engineered nucleases such as zinc finger nucleases or engineered homing endonucleases.

Genetic engineering techniques have been applied in numerous fields including research, biotechnology, and medicine. Medicines such as insulin and human growth hormone are now produced in bacteria, experimental mice such as the oncomouse and the knockout mouse are being used for research purposes and insect resistant and/or herbicide tolerant crops have been commercialized. Genetically engineered plants and animals capable of producing biotechnology drugs more cheaply than current methods (called pharming) are also being developed and in 2009 the FDA approved the sale of the pharmaceutical protein antithrombin produced in the milk of genetically engineered goats.

Definition

Genetic engineering alters the genetic makeup of an organism using techniques that introduce heritable material prepared outside the organism either directly into the host or into a cell that is then fused or hybridized with the host. This involves using recombinant nucleic acid (DNA or RNA) techniques to form new combinations of heritable genetic material followed by the incorporation of that material either indirectly through a vector system or directly through micro-injection, macro-injection and micro-encapsulation techniques. Genetic engineering does not include traditional animal and plant breeding, in vitro fertilisation, induction of polyploidy, mutagenesis and cell fusion techniques that do not use recombinant nucleic acids or a genetically modified organism in the process. Cloning and stem cell research, although not considered genetic engineering, are closely related and genetic engineering can be used within them. Synthetic biology is an emerging discipline that takes genetic engineering a step further by introducing artificially synthesized genetic material from raw materials into an organism. If genetic material from another species is added to the host, the resulting organism is called transgenic. If genetic material from the same species or a species that can naturally breed with the host is used the resulting organism is called cisgenic. Genetic engineering can also be used to remove genetic material from the target organism, creating a knock out organism. In Europe genetic modification is synonymous with genetic engineering while within the United States of America it can also refer to conventional breeding methods.

History

Humans have altered the genomes of species for thousands of years through artificial selection and more recently mutagenesis. Genetic engineering as the direct manipulation of DNA by humans outside breeding and mutations has only existed since the 1970s. The term "genetic engineering" was first coined by Jack Williamson in his science fiction novel *Dragon's Island*, published in 1951, one year before DNA's role in heredity was confirmed by Alfred Hershey and Martha Chase, and two years before James Watson and Francis Crick showed that the DNA molecule has a double-helix structure.

In 1972 Paul Berg created the first recombinant DNA molecules by combined DNA from the monkey virus SV40 with that of the lambda virus. In 1973 Herbert Boyer and Stanley Cohen created the first transgenic organism by inserting antibiotic resistance genes into the plasmid of an *E. coli* bacterium. A year later Rudolf Jaenisch created a transgenic mouse by introducing foreign DNA into its embryo, making it the world's first transgenic animal. In 1976 Genentech, the first genetic engineering company was founded by Herbert Boyer and Robert Swanson and a year later and the company produced a human protein (somatostatin) in *E.coli*. Genentech announced the production of genetically engineered human insulin in 1978. In 1980, the U.S. Supreme Court in the Diamond v. Chakrabarty case ruled that genetically altered life could be patented. The insulin produced by bacteria, branded humulin, was approved for release by the Food and Drug Administration in 1982.

The first field trials of genetically engineered plants occurred in France and the USA in 1986, tobacco plants were engineered to be resistant to herbicides. The People's Republic of China was the first country to commercialize transgenic plants, introducing a virus-resistant tobacco in 1992. In 1994 Calgene attained approval to commercially release the Flavr Savr tomato, a tomato engineered to have a longer shelf life. In 1994, the European Union approved tobacco engineered to be resistant to the herbicide bromoxynil, making it the first genetically engineered crop commercialized in Europe. In 1995, Bt Potato was approved safe by the Environmental Protection Agency, making it the first pesticide producing crop to be approved in the USA. In 2009 11 transgenic crops were grown commercially in 25 countries, the largest of which by area grown were the USA, Brazil, Argentina, India, Canada, China, Paraguay and South Africa.

In 2010, scientists at the J. Craig Venter Institute, announced that they had created the first synthetic bacterial genome, and added it to a cell containing no DNA. The resulting bacterium, named Synthia, was the world's first synthetic life form.

Process

Isolating the Gene: First, the gene to be inserted into the genetically modified organism must be chosen and isolated. Presently, most genes transferred into plants provide protection against insects or tolerance to herbicides. In animals the majority of genes used are growth hormone genes. Once chosen the genes must be isolated. This typically involves multiplying the gene using polymerase chain reaction (PCR). If the chosen gene or the donor organism's genome has been well studied it may be present in a genetic library. If the DNA sequence is known, but no copies of the gene are available, it can be artificially synthesized. Once isolated, the gene is inserted into a bacterial plasmid.

Constructs: The gene to be inserted into the genetically modified organism must be combined with other genetic elements in order for it to work properly. The gene can also be modified at this stage for better expression or effectiveness. As well as the gene to be inserted most constructs contain a promoter and terminator region as well as a selectable marker gene. The promoter region initiates transcription of the gene and can be used to control the location and level of gene expression, while the terminator region ends transcription. The selectable marker, which in most cases confers antibiotic resistance to the organism it is expressed in, is needed to determine which cells are transformed with the new gene. The constructs are made using recombinant DNA techniques, such as restriction digests, ligations and molecular cloning.

Gene Targeting: The most common form of genetic engineering involves inserting new genetic material randomly within the host genome. Other techniques allow new genetic material to be inserted at a specific location in the host genome or generate mutations at desired genomic loci capable of knocking out endogenous genes. The technique of gene targeting uses homologous recombination to target desired changes to a specific endogenous gene. This tends to occur at a relatively low frequency in plants and animals and generally requires the use of selectable markers. The frequency of gene targeting can be greatly enhanced with the use of engineered nucleases such as zinc finger nucleases, engineered homing endonucleases, or nucleases created from TAL effectors. In addition to enhancing gene targeting, engineered nucleases can also be used to introduce mutations at endogenous genes that generate a gene knockout.

Transformation: About 1% of bacteria are naturally able to take up foreign DNA but it can also be induced in other bacteria. Stressing the bacteria for example, with a heat shock or an electric shock, can make the cell membrane permeable to DNA that may then incorporate into their genome or exist as extrachromosomal DNA. DNA is generally inserted into animal cells using microinjection, where it can be injected through the cells nuclear envelope directly into the nucleus or through the use of viral vectors. In plants the DNA is generally inserted using *Agrobacterium*-mediated recombination or biolistics.

In *Agrobacterium*-mediated recombination the plasmid construct must also contain T-DNA. *Agrobacterium* naturally inserts DNA from a tumor inducing plasmid into any susceptible plant's genome it infects, causing crown gall disease. The T-DNA region of this plasmid is responsible for insertion of the DNA. The genes to be inserted are cloned into a binary vector, which contains T-DNA and can be grown in both *E. Coli* and *Agrobacterium*. Once the binary vector is constructed the plasmid is transformed into *Agrobacterium* containing no plasmids and plant cells are infected. The *Agrobacterium* will then naturally insert the genetic material into the plant cells.

In biolistics particles of gold or tungsten are coated with DNA and then shot into young plant cells or plant embryos. Some genetic material will enter the cells and transform them. This method can be used on plants that are not susceptible to *Agrobacterium* infection and also allows transformation of plant plastids.

Another transformation method for plant and animal cells is electroporation. Electroporation involves subjecting the plant or animal cell to an electric shock, which can make the cell membrane permeable to plasmid DNA. In some cases the electroporated cells will incorporate the DNA into their genome. Due to the damage caused to the cells and DNA the transformation efficiency of biolistics and electroporation is lower than agrobacterial mediated transformation and microinjection.

Selection: Not all the organism's cells will be transformed with the new genetic material; in most cases a selectable marker is used to differentiate transformed from untransformed cells. If a cell has been successfully transformed with the DNA it will also contain the marker gene.

By growing the cells in the presence of an antibiotic or chemical that selects or marks the cells expressing that gene it is possible to separate the transgenic events from the non-transgenic. Another method of screening involves using a DNA probe that will only stick to the inserted gene. A number of strategies have been developed that can remove the selectable marker from the mature transgenic plant.

Regeneration: As often only a single cell is transformed with genetic material the organism must be regrown from that single cell. As bacteria consist of a single cell and reproduce clonally regeneration is not necessary. In plants this is accomplished through the use of tissue culture.

Each plant species has different requirements for successful regeneration through tissue culture. If successful an adult plant is produced that contains the transgene in every cell. In animals it is necessary to ensure that the inserted DNA is present in the embryonic stem cells. When the offspring is produced they can be screened for the presence of the gene. All offspring from the first generation will be heterozygous for the inserted gene and must be mated together to produce a homozygous animal.

Confirmation: Further tests using PCR, Southern Blots and Bioassays are needed to confirm that the gene is expressed and functions correctly. The organism's offspring are also tested to ensure that the trait can be inherited and that it follows a Mendelian inheritance pattern.

Applications

Genetic engineering has applications in medicine, research, industry and agriculture and can be used on a wide range of plants, animals and micro organism.

Medicine: In medicine genetic engineering has been used to mass-produce insulin, human growth hormones, follistim (for treating infertility), human albumin, monoclonal antibodies, antihemophilic factors, vaccines and many other drugs. Vaccination generally involves injecting weak live, killed or inactivated forms of viruses or their toxins into the person being immunized. Genetically engineered viruses are being developed that can still confer immunity, but lack the infectious sequences. Mouse hybridomas, cells fused together to create monoclonal antibodies, have been humanised through genetic engineering to create human monoclonal antibodies.

Genetic engineering is used to create animal models of human diseases. Genetically modified mice are the most common genetically engineered animal model. They have been used to study and model cancer (the oncomouse), obesity, heart disease, diabetes, arthritis, substance abus, anxiety, aging and Parkinson disease. Potential cures can be tested against these mouse models. Also genetically modified pigs have been bred with the aim of increasing the success of pig to human organ transplantation.

Gene therapy is the genetic engineering of humans by replacing defective human genes with functional copies. This can occur in somatic tissue or germline tissue. If the gene is inserted into the germline tissue it can be passed down to that person's descendants. Gene therapy has been used to treat patients suffering from immune deficiencies (notably Severe combined immunodeficiency) and trials have been carried out on other genetic disorders. The success of gene therapy so far has been limited and a patient (Jesse Gelsinger) has died during a clinical trial testing a new treatment. There are also ethical concerns should the technology be used not just for treatment, but for enhancement, modification or alteration of a human beings' appearance, adaptability, intelligence, character or behaviour. The distinction between cure and enhancement can also be difficult to establish. Transhumanists consider the enhancement of humans desirable.

Research: Genetic engineering is an important tool for natural scientists. Genes and other genetic information from a wide range of organisms are transformed into bacteria for storage and modification, creating genetically modified bacteria in the process. Bacteria are cheap, easy to grow, clonal, multiply quickly, relatively easy to transform and can be stored at -80°C almost indefinitely. Once a gene is isolated it can be stored inside the bacteria providing an unlimited supply for research. Organisms are genetically engineered to discover the functions of certain genes. This could be the effect on the phenotype of the organism, where the gene is expressed or what other genes it interacts with. These experiments generally involve loss of function, gain of function, tracking and expression.

- Loss of function experiments, such as in a gene knockout experiment, in which an organism is engineered to lack the activity of one or more genes. A knockout experiment involves the creation and manipulation of a DNA construct *in vitro*, which, in a simple knockout, consists of a copy of the desired gene, which has been altered such that it is non-functional. Embryonic stem cells incorporate the altered gene, which replaces the already present functional copy. These stem cells are injected into blastocysts, which are implanted into surrogate mothers. This allows the experimenter to analyze the defects caused by this mutation and thereby determine the role of particular genes. It is used especially frequently in developmental biology. Another method, useful in organisms such as Drosophila (fruit fly), is to induce mutations in a large population and then screen the progeny for the desired mutation. A similar process can be used in both plants and prokaryotes.
- Gain of function experiments, the logical counterpart of knockouts. These are sometimes performed in conjunction with knockout experiments to more finely establish the function of the desired gene. The process is much the same as that in knockout engineering, except that the construct is designed to increase the function of the gene, usually by providing extra copies of the gene or inducing synthesis of the protein more frequently.
- Tracking experiments, which seek to gain information about the localization and interaction of the desired protein. One way to do this is to replace the wild-type gene with a 'fusion' gene, which is a juxtaposition of the wild-type gene with a reporting element such as green fluorescent protein (GFP) that will allow easy visualization of the products of the genetic modification. While this is a useful technique, the manipulation can destroy the function of the gene, creating secondary effects

and possibly calling into question the results of the experiment. More sophisticated techniques are now in development that can track protein products without mitigating their function, such as the addition of small sequences that will serve as binding motifs to monoclonal antibodies.

- Expression studies aim to discover where and when specific proteins are produced. In these experiments, the DNA sequence before the DNA that codes for a protein, known as a gene's promoter, is reintroduced into an organism with the protein coding region replaced by a reporter gene such as GFP or an enzyme that catalyzes the production of a dye. Thus the time and place where a particular protein is produced can be observed. Expression studies can be taken a step further by altering the promoter to find which pieces are crucial for the proper expression of the gene and are actually bound by transcription factor proteins; this process is known as promoter bashing.

Industrial: By engineering genes into bacterial plasmids it is possible to create a biological factory that can produce proteins and enzymes. Some genes do not work well in bacteria, so yeast, a eukaryote, can also be used. Bacteria and yeast factories have been used to produce medicines such as insulin, human growth hormone, and vaccines, supplements such as tryptophan, aid in the production of food (chymosin in cheese making) and fuels. Other applications involving genetically engineered bacteria being investigated involve making the bacteria perform tasks outside their natural cycle, such as cleaning up oil spills, carbon and other toxic waste.

Agriculture: One of the best-known and controversial applications of genetic engineering is the creation of genetically modified food. There are three generations of genetically modified crops. First generation crops have been commercialized and most provide protection from insects and/or resistance to herbicides. There are also fungal and virus resistant crops developed or in development. They have been developed to make the insect and weed management of crops easier and can indirectly increase crop yield. The second generation of genetically modified crops being developed aim to directly improve yield by improving salt, cold or drought tolerance and to increase the nutritional value of the crops. The third generation consists of pharmaceutical crops, crops that contain edible vaccines and other drugs. Some agriculturally important animals have been genetically modified with growth hormones to increase their size while others have been engineered to express drugs and other proteins in their milk. The genetic engineering of agricultural crops can increase the growth rates and resistance to different diseases caused by pathogens and parasites. This is beneficial as it can greatly increase the production of food sources with the usage of fewer resources that would be required to host the world's growing populations. These modified crops would also reduce the usage of chemicals, such as fertilizers and pesticides, and therefore decrease the severity and frequency of the damages produced by these chemical pollution. Ethical and safety concerns have been raised around the use of genetically modified food. A major safety concern relates to the human health implications of eating genetically modified food, in particular whether toxic or allergic reactions could occur. Gene flow into related non-transgenic crops, off target effects on beneficial organisms and the impact on biodiversity are important environmental issues. Ethical concerns involve religious issues, corporate control of the food supply, intellectual property rights and the level of labeling needed on genetically modified products.

Other Uses: In materials science, a genetically modified virus has been used to construct a more environmentally friendly lithium-ion battery. Some bacteria have been genetically engineered to create black and white photographs while others have potential to be used as sensors by expressing a fluorescent protein under certain environmental conditions. Genetic engineering is also being used to create BioArt and novelty items such as blue roses, and glowing fish.

Opposition and Criticism

A 2010 study of Canola found transgenes in 80% of wild (uncultivated or "feral") varieties in North Dakota, meaning 80% of the plants which had established themselves in the area were genetically engineered varieties. The researchers stated that "we found the highest densities of [such transgene-containing] plants near agricultural fields and along major freeways, but we were also finding plants in the middle of nowhere" adding that "over time,..the build-up of different types of herbicide resistance in feral [natural] canola and closely related weeds, like field mustard, could make it more difficult to manage these plants using herbicides."

Ethanol Fermentation

Ethanol fermentation, also referred to as alcoholic fermentation, is a biological process in which sugars such as glucose, fructose, and sucrose are converted into cellular energy and thereby produce ethanol and carbon dioxide as metabolic waste products. Because yeasts perform this conversion in the absence of oxygen, ethanol fermentation is classified as anaerobic.

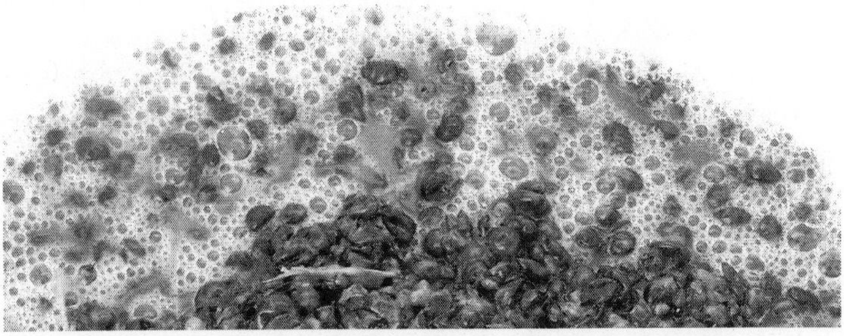

Figure 7: Grapes fermenting during the production of wine.

Ethanol fermentation occurs in the production of alcoholic beverages and ethanol fuel, and in the rising of bread dough.

The Chemical Process of Fermentation of Glucose

The chemical equation below summarizes the fermentation of glucose, whose chemical formula is $C_6H_{12}O_6$. One mole of glucose is converted into two moles of ethanol and two moles of carbon dioxide:

$$C_{12}H_{22}O_{11} + H_2O + \text{invertase} \rightarrow C_6H_{12}O_6$$

$$C_6H_{12}O_6 + \text{Zymase} \rightarrow 2C_2H_5OH + 2CO_2$$

C_2H_5OH is the chemical formula for ethanol.

Before fermentation takes place, one glucose molecule is broken down into two pyruvate molecules. This is known as glycolysis. Glycolysis is summarized by the chemical equation:

$$C_6H_{12}O_6 + 2\ ADP + 2\ P_i + 2\ NAD^+ \rightarrow 2\ CH_3COCOO^- + 2\ ATP + 2\ NADH + 2\ H_2O + 2H^+$$

The chemical formula of pyruvate is CH_3COCOO^-. P_i stands for the inorganic phosphate. As shown by the reaction equation, glycolysis causes the reduction of two molecules of NAD^+ to NADH. Two ADP molecules are also converted to two ATP and two water molecules via substrate-level phosphorylation.

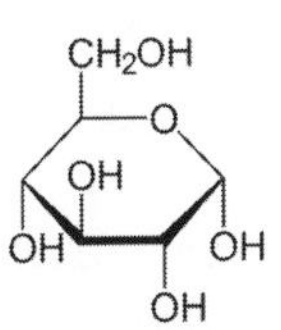

Glucose depicted in Haworth projection

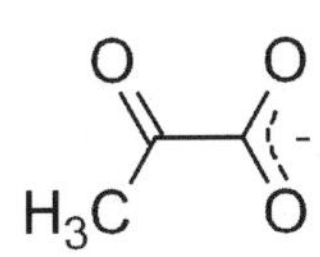

Pyruvate

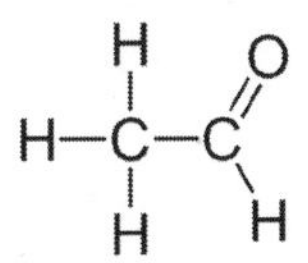

Acetaldehyde

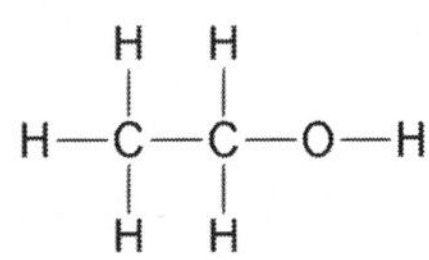

Ethanol

Effect of Oxygen

The fermentation process does not require oxygen. If oxygen is present, some species of yeast (*Kluyveromyces lactis*, *Kluyveromyces lipolytica*) oxidize pyruvate completely to carbon dioxide and water. This process is called respiration. Thus these yeasts produce ethanol only in an anaerobic environment. However, many yeasts such as the commonly used baker's yeast *Saccharomyces cerevisiae*, and *Schizosaccharomyces pombe*, prefer fermentation to respiration. These yeasts will produce ethanol even under aerobic conditions given the right sources of nutrition.

Uses

Ethanol fermentation is responsible for the rising of bread dough. Yeast organisms consume sugars in the dough and produce ethanol and carbon dioxide as waste products. The carbon dioxide forms bubbles in the dough, expanding it into something of a foam. Nearly all the ethanol evaporates from the dough when the bread is baked.

All alcoholic beverages, including those produced by carbonic maceration, are produced by ethanol fermentation by yeast. Wine and brandy are produced by fermentation of the natural sugars present in fruits, especially grapes. Beer and whiskey are produced by fermentation of grain starches that have been converted to sugar by the enzyme amylase, which is present in grain kernels that have been germinated. Amylase-treated grain or amylase-treated potatoes are fermented for the production of vodka. Rum is produced by fermentation of cane sugar. In all cases, the fermentation must take place in a vessel that allows carbon dioxide to escape, but prevents outside air from coming in, as exposure to oxygen would prevent the formation of ethanol.

Similar yeast fermentation of various carbohydrate products is used to produce much of the ethanol used for fuel.

Feedstocks for Fuel Production

The dominant ethanol feedstock in warmer regions is sugarcane. In temperate regions, sugar beet is sometimes used instead. In the United States, the main feedstock for the production of ethanol is currently corn. Approximately 2.8 gallons of ethanol are produced from one bushel of corn (0.42 liter per kilogram). While much of the corn turns into ethanol, some of the corn also yields by-products such as DDGS (distillers dried grains with solubles) that can be used as feed for livestock. A bushel of corn produces about 18 pounds of DDGS

(320 kilograms of DDGS per metric ton of maize). Although most of the fermentation plants have been built in corn-producing regions, sorghum is also an important feedstock for ethanol production in the Plains states. Pearl millet is showing promise as an ethanol feedstock for the southeastern U.S. and the potential of duckweed is being studied.

In some parts of Europe, particularly France and Italy, grapes have become a *de facto* feedstock for fuel ethanol by the distillation of surplus wine. In Japan, it has been proposed to use rice normally made into sake as an ethanol source.

Cassava as Ethanol Feedstock: Ethanol can be made from mineral oil or from sugars or starches. Starches are cheapest. The starchy crop with highest energy content per acre is cassava, which grows in tropical countries. Thailand already had a large cassava industry in the 1990s, for use as cattle feed and as a cheap admixture to wheat flour. Nigeria and Ghana are already establishing cassava-to-ethanol plants. Production of ethanol from cassava is currently economically feasible when crude oil prices are above US$120 per barrel. New varieties of cassava are being developed, so the future situation remains uncertain. Currently, cassava can yield between 25-40 tonnes per hectare (with irrigation and fertilizer), and from a tonne of cassava roots, circa 200 liters of ethanol can be produced (assuming cassava with 22% starch content). A liter of ethanol contains circa 21.46 MJ of energy. The overall energy efficiency of cassava-root to ethanol conversion is circa 32%. The yeast used for processing cassava is *Endomycopsis fibuligera*, sometimes used together with bacterium *Zymomonas mobilis*.

Byproducts of Fermentation

Ethanol fermentation produces unharvested byproducts such as heat, food for livestock, and water.

Microbes Used in Ethanol Fermentation

- Yeast
- Zymomonas mobilis

Chapter 4

Testing of DNA and RNA

The nucleic acids are the building blocks of living organisms. You may have heard of DNA described the same way. Guess what? DNA is just one type of nucleic acid. Some other types are RNA, mRNA, and tRNA. All of these “Na’s” work together to help cells replicate and build proteins.

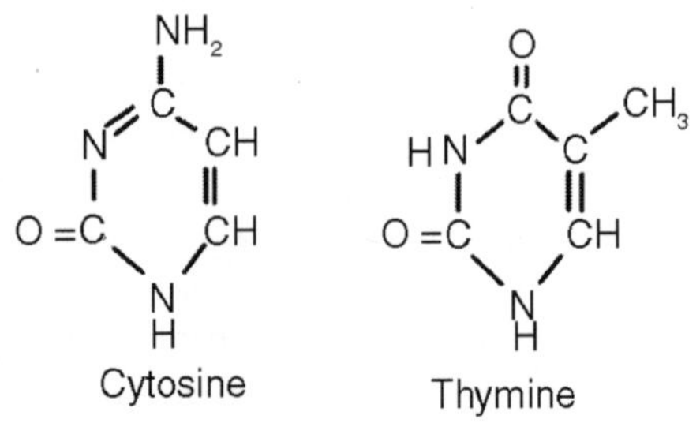

Pyrimidine Bases

Basics

We already told you about the biggie nucleic acids (DNA, RNA, mRNA, tRNA). They are actually made up chains of base pairs stretching from only a few to millions. When those pairs combine in super long chains (DNA), they make a shape called a double helix. The double helix shape is like a twisty ladder. The base pairs are the rungs. We’re very close to talking about the biology of cells here. Back to the chemistry..

Five Easy Pieces

There are five easy parts of nucleic acids. All nucleic acids are made up of the same building blocks (monomers). Chemists call the monomers nucleotides. The five pieces are Uracil, Cytosine, Thymine, Adenine, and Guanine. Just as there are twenty (20) amino acids needed by humans to survive, there are five (5)nucleotides.

These nucleotides are made of three parts:

1. A five carbon sugar
2. A base that has a nitrogen (N) atom
3. An ion of phosphoric acid

The first isolation of what we now refer to as DNA was accomplished by Johann Friedrich Miescher *circa* 1870. He reported finding a weakly acidic substance of unknown function in the nuclei of human white blood cells, and named this material "nuclein". A few years later, Miescher separated nuclein into protein and nucleic acid components. In the 1920's nucleic acids were found to be major components of chromosomes, small gene-carrying bodies in the nuclei of complex cells. Elemental analysis of nucleic acids showed the presence of phosphorus, in addition to the usual C, H, N & O. Unlike proteins, nucleic acids contained no sulphur. Complete hydrolysis of chromosomal nucleic acids gave inorganic phosphate, 2-deoxyribose (a previously unknown sugar) and four different heterocyclic bases (shown in the following diagram). To reflect the unusual sugar component, chromosomal nucleic acids are called deoxyribonucleic acids, abbreviated DNA. Analogous nucleic acids in which the sugar component is ribose are termed ribonucleic acids, abbreviated RNA. The acidic character of the nucleic acids was attributed to the phosphoric acid moiety.

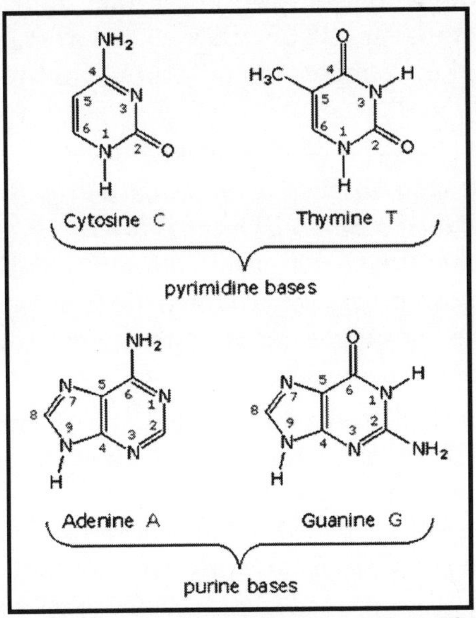

The two monocyclic bases shown here are classified as pyrimidines, and the two bicyclic bases are purines. Each has at least one N-H site at which an organic substituent may be attached. They are all polyfunctional bases, and may exist in tautomeric forms.

Base-catalyzed hydrolysis of DNA gave four nucleoside products, which proved to be N-glycosides of 2'-deoxyribose combined with the heterocyclic amines. Structures and names for these nucleosides will be displayed above by clicking on the heterocyclic base diagram. The base components are coloured green, and the sugar is black. As noted in the 2'-deoxycytidine structure on the left, the numbering of the sugar carbons makes use of primed numbers to distinguish them from the heterocyclic base sites. The corresponding N-glycosides of the common sugar ribose are the building blocks of RNA, and are named adenosine, cytidine, guanosine and uridine (a thymidine analog missing the methyl group).

From this evidence, nucleic acids may be formulated as alternating copolymers of phosphoric acid (P) and nucleosides (N), as shown:

$$\sim P - N - P - N' - P - N'' - P - N''' - P - N \sim$$

At first the four nucleosides, distinguished by prime marks in this crude formula, were assumed to be present in equal amounts, resulting in a uniform structure, such as that of starch. However, a compound of this kind, presumably common to all organisms, was considered too simple to hold the hereditary information known to reside in the chromosomes. This view was challenged in 1944, when Oswald Avery and colleagues demonstrated that bacterial DNA was likely the geneic agent that carried information from one organism to another in a process called "transformation".

He concluded that *"nucleic acids must be regarded as possessing biological specificity, the chemical basis of which is as yet undetermined."* Despite this finding, many scientists continued to believe that chromosomal proteins, which differ across species, between individuals, and even within a given organism, were the locus of an organism's genetic information.

It should be noted that single celled organisms like bacteria do not have a well-defined nucleus. Instead, their single chromosome is associated with specific roteins in a region called a "nucleoid". Nevertheless, the DNA from bacteria has the same composition and general structure as that from multicellular organisms, including human beings.

Views about the role of DNA in inheritance changed in the late 1940's and early 1950's. By conducting a careful analysis of DNA from many sources, Erwin Chargaff found its composition to be species specific. In addition, he found that the amount of adenine (A) always equaled the amount of thymine (T), and the amount of guanine (G) always equaled the amount of cytosine (C), regardless of the DNA source. As set forth in the following table, the ratio of (A+T) to (C+G) varied from 2.70 to 0.35. The last two organisms are bacteria.

Nucleoside Base Distribution in DNA

Organism	Base Composition (mole %)				Base Ratios		Ratio (A+T)/(G+C)
	A	G	T	C	A/T	G/C	
Human	30.9	19.9	29.4	19.8	1.05	1.00	1.52
Chicken	28.8	20.5	29.2	21.5	1.02	0.95	1.38
Yeast	31.3	18.7	32.9	17.1	0.95	1.09	1.79
Clostridium perfringens	36.9	14.0	36.3	12.8	1.01	1.09	2.70
Sarcina lutea	13.4	37.1	12.4	37.1	1.08	1.00	0.35

In a second critical study, Alfred Hershey and Martha Chase showed that when a bacterium is infected and genetically transformed by a virus, at least 80% of the viral DNA enters the bacterial cell and at least 80% of the viral protein remains outside. Together with the Chargaff findings this work established DNA as the repository of the unique genetic characteristics of an organism.

The Chemical Nature of DNA

The polymeric structure of DNA may be described in terms of monomeric units of increasing complexity. In the top shaded box of the following illustration, the three relatively simple components mentioned earlier are shown. Below that on the left, formulas for phosphoric acid and a nucleoside are drawn.Condensation polymerization of these leads to the DNA formulation outlined above. Finally, a 5'- monophosphate ester, called a nucleo-tide may be drawn as a single monomer unit, shown in the shaded box to the right. Since a monophosphate ester of this kind is a strong acid (pK_a of 1.0), it will be fully ionized at the usual physiological pH (ca.7.4).

Names for these DNA components are given in the table to the right of the diagram. Isomeric 3'-monophospate nucleotides are also known, and both isomers are found in cells. They may be obtained by selective hydrolysis of DNA through the action of nuclease enzymes.

Anhydride-like di- and tri-phosphate nucleotides have been identified as important energy carriers in biochemical reactions, the most common being ATP (adenosine 5'-triphosphate).

Names of DNA Base Derivatives

Base	Nucleoside	5'-Nucleotide
Adenine	2'-Deoxyadenosine	2'-Deoxyadenosine-5'-monophosphate
Cytosine	2'-Deoxycytidine	2'-Deoxycytidine-5'-monophosphate
Guanine	2'-Deoxyguanosine	2'-Deoxyguanosine-5'-monophosphate
Thymine	2'-Deoxythymidine	2'-Deoxythymidine-5'-monophosphate

A complete structural representation of a segment of the DNA polymer formed from 5'-nucleotides. Several important characteristics of this formula should be noted.

- First, the remaining P-OH function is quite acidic and is completely ionized in biological systems.
- Second, the polymer chain is structurally directed. One end is different from the other.
- Third, although this appears to be a relatively simple polymer, the possible permutations of the four nucleosides in the chain become very large as the chain lengthens.
- Fourth, the DNA polymer is much larger than originally believed. Molecular weights for the DNA from multicellular organisms are commonly 10^9 or greater.

Information is stored or encoded in the DNA polymer by the pattern in which the four nucleotides are arranged. To access this information the pattern must be "read" in a linear fashion, just as a bar code is read at a supermarket checkout. Because living organisms are extremely complex, a correspondingly large amount of information related to this complexity must be stored in the DNA.

Consequently, the DNA itself must be very large, as noted above. Even the single DNA molecule from an *E. coli* bacterium is found to have roughly a million nucleotide units in a polymer strand, and would reach a millimeter in length if stretched out. The nuclei of multicellular organisms incorporate chromosomes, which are composed of DNA combined with nuclear proteins called histones. The fruit fly has 8 chromosomes, humans have 46 and dogs 78 (note that the amount of DNA in a cell's nucleus does not correlate with the number of chromosomes). The DNA from the smallest human chromosome is over ten times larger than *E. coli* DNA, and it has been estimated that the total DNA in a human cell would extend to 2 meters in length if unraveled. Since the nucleus is only about 5ìm in diameter, the chromosomal DNA must be packed tightly to fit in that small volume.

In addition to its role as a stable informational library, chromosomal DNA must be structured or organized in such a way that the chemical machinery of the cell will have easy access to that information, in order to make important molecules such as polypeptides. Furthermore, accurate copies of the DNA code must be created as cells divide, with the replicated DNA molecules passed on to subsequent cell generations, as well as to progeny of the organism.

RNA, a Different Nucleic Acid

The high molecular weight nucleic acid, DNA, is found chiefly in the nuclei of complex cells, known as eucaryotic cells, or in the nucleoid regions ofprocaryotic cells, such as bacteria. It is often associated with proteins that help to pack it in a usable fashion. In contrast, a lower molecular weight, but much more abundant nucleic acid, RNA, is distributed throughout the cell, most commonly in small numerous organelles called ribosomes. Three kinds of RNA are identified, the largest subgroup (85 to 90%) being ribosomal RNA, rRNA, the major component of ribosomes, together with proteins. The size of rRNA molecules varies, but is generally less than a thousandth the size of DNA. The other forms of RNA are messenger RNA, mRNA, and transfer RNA, tRNA. Both have a more transient existence and are smaller than rRNA.

All these RNA's have similar constitutions, and differ from DNA in two important respects. As shown in the following diagram, the sugar component of RNA is ribose, and the pyrimidine base uracil replaces the thymine base of DNA. The RNA's play a vital role in the transfer of information (transcription) from the DNA library to the protein factories called ribosomes, and in the interpretation of that information (translation) for the synthesis of specific polypeptides. These functions will be described later. A complete structural representation of a segment of the RNA polymer formed from 5'-nucleotides.

The Secondary Structure of DNA

In the early 1950's the primary structure of DNA was well established, but a firm understanding of its secondary structure was lacking. Indeed, the situation was similar to that occupied by the proteins a decade earlier, before the alpha helix and pleated sheet structures were proposed by Linus Pauling. Many researchers grappled with this problem, and it was generally conceded that the molar equivalences of base pairs (A & T and C & G) discovered by Chargaff would be an important factor. A scientist, working at King's College, obtained X-ray diffraction evidence that suggested a long helical structure of uniform thickness.

The Double Helix

After many trials and modifications, an ingenious double helix model for the secondary structure of DNA. Two strands of DNA were aligned anti-parallel to each other, *i.e.* with opposite 3' and 5' ends, as shown in part a of the following diagram. Complementary primary nucleotide structures for each strand allowed intra-strand hydrogen bonding between each pair of bases. These complementary strands are coloured red and green in the diagram. Coiling these coupled strands then leads to a double helix structure, shown as cross-linked ribbons in part b of the diagram. A space-filling molecular model of a short segment is displayed in part c on the right.

The helix shown here has ten base pairs per turn, and rises 3.4 Å in each turn. This right-handed helix is the favoured conformation in aqueous systems, and has been termed the B-helix. Two alternating grooves are evident, a wide and deep major groove, and a shallow and narrow minor groove. Other molecules, including polypeptides, may insert into these grooves, and in so doing perturb the chemistry of DNA. Other helical structures of DNA have also been observed, and are designated by letters (*e.g.* A and Z).

The Double Helix Structure for DNA

DNA Replication: In their announcement of a double helix structure for DNA, It is stated, *"It has not escaped our notice that the specific pairing we have postulated immediately suggests a possible copying mechanism for the genetic material."*. The essence of this suggestion is that, if separated, each strand of the molecule might act as a template on which a new complementary strand might be assembled, leading finally to two identical DNA molecules. Indeed, replication does take place in this fashion when cells divide, but the events leading up to the actual synthesis of complementary DNA strands are sufficiently complex that they will not be described in any detail.

As depicted in the following drawing, the DNA of a cell is tightly packed into chromosomes. First, the DNA is wrapped around small proteins called histones. These bead-like structures are then further organized and folded into chromatin aggregates that make up the chromosomes.

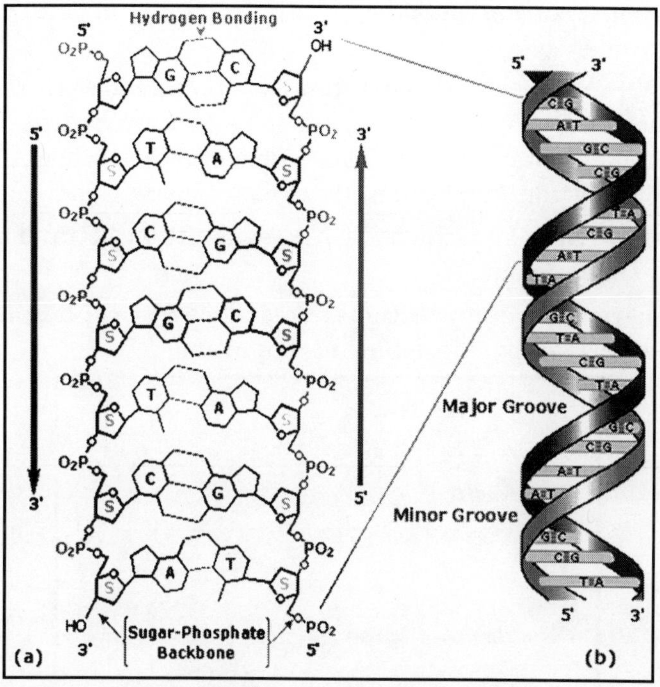

An overall packing efficiency of 7,000 or more is thus achieved. Clearly a sequence of unfolding events must take place before the information encoded in the DNA can be used or replicated.

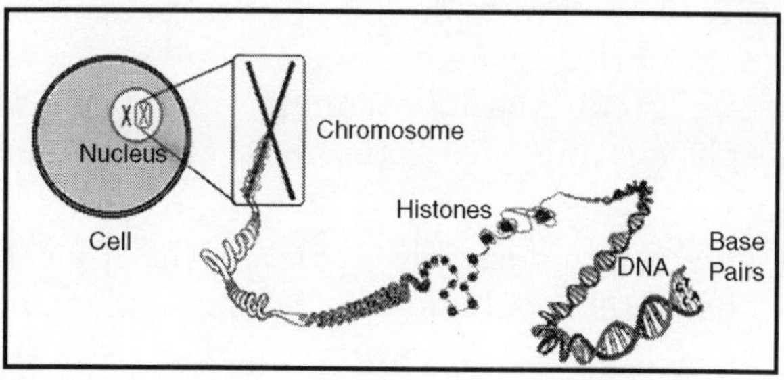

Once the double stranded DNA is exposed, a group of enzymes act to accomplish its replication.

These are described briefly here:

- *Topoisomerase:* This enzyme initiates unwinding of the double helix by cutting one of the strands.
- *Helicase:* This enzyme assists the unwinding. Note that many hydrogen bonds must be broken if the strands are to be separated.

- *SSB:* A single-strand binding-protein stabilizes the separated strands, and prevents them from recombining, so that the polymerization chemistry can function on the individual strands.
- *DNA Polymerase:* This family of enzymes link together nucleotide triphosphate monomers as they hydrogen bond to complementary bases. These enzymes also check for errors (roughly ten per billion), and make corrections.
- *Ligase:* Small unattached DNA segments on a strand are united by this enzyme.

Polymerization of nucleotides takes place by the phosphorylation reaction described by the following equation.

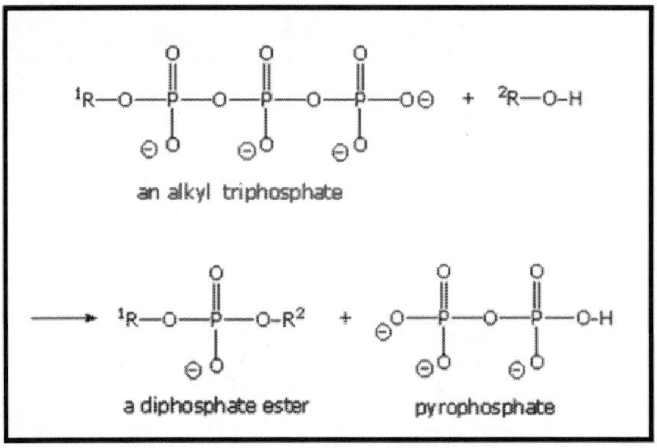

Di- and triphosphate esters have anhydride-like structures and are consequently reactive phosphorylating reagents, just as carboxylic anhydrides are acylating reagents. Since the pyrophosphate anion is a better leaving group than phosphate, triphosphates are more powerful phosphorylating agents than are diphosphates.

Formulas for the corresponding 5'-derivatives of adenosine will be displayed, and similar derivatives exist for the other three common nucleosides. The DNA polymerization process that builds the complementary strands in replication, could in principle take place in two ways. Referring to the general equation above, R^1 could represent the next nucleotide unit to be attached to the growing DNA strand, with R^2 being this strand. Alternatively, these assignments could be reversed.

In practice, the former proves to be the best arrangement. Since triphosphates are very reactive, the lifetime of such derivatives in an aqueous environment is relatively short. However, such derivatives of the individual nucleosides are repeatedly synthesized by the cell for a variety of purposes, providing a steady supply of these reagents.

In contrast, the growing DNA segment must maintain its functionality over the entire replication process, and can not afford to be changed by a spontaneous hydrolysis event. As a result, these chemical properties are best accomodated by a polymerization process that proceeds at the 3'-end of the growing strand by 5'-phosphorylation involving a nucleotide triphosphate.

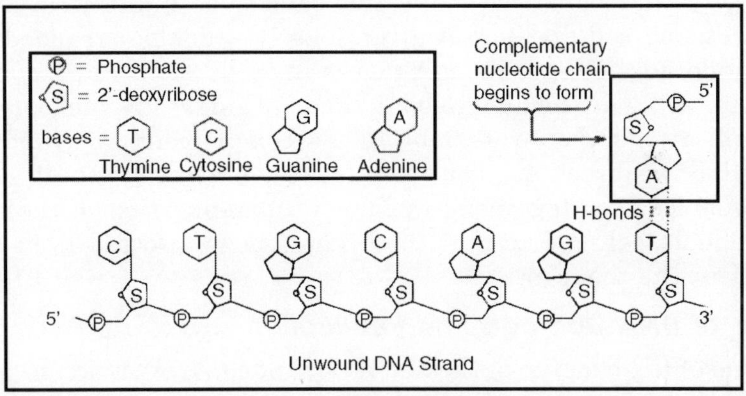

The polymerization mechanism described here is constant. It always extends the developing DNA segment towards the 3'-end (*i.e.* when a nucleotide triphosphate attaches to the free 3'-hydroxyl group of the strand, a new 3'-hydroxyl is generated). There is sometimes confusion on this point, because the original DNA strand that serves as a template is read from the 3'-end towards the 5'-end, and authors may not be completely clear as to which terminology is used.

Because of the directional demand of the polymerization, one of the DNA strands is easily replicated in a continuous fashion, whereas the other strand can only be replicated in short segmental pieces. This is illustrated in the following diagram. Separation of a portion of the double helix takes place at a site called the replication fork. As replication of the separate strands occurs, the replication fork moves away (to the left in the diagram), unwinding additional lengths of DNA. Since the fork in the diagram is moving towards the 5'-end of the red-coloured strand, replication of this strand may take place in a continuous fashion (building the new green strand in a 5' to 3' direction). This continuously formed new strand is called the leading strand. In contrast, the replication fork moves towards the 3'-end of the original green strand, preventing continuous polymerization of a complementary new red strand. Short segments of complementary DNA, called Okazaki fragments, are produced, and these are linked together later by the enzyme ligase. This new DNA strand is called the lagging strand.

When you consider that a human cell has roughly 10^9 base pairs in its DNA, and may divide into identical daughter cells in 14 to 24 hours, the efficiency of DNA replication must be extraordinary. The procedure described above will replicate about 50 nucleotides per second, so there must be many thousand such replication sites in action during cell division. A given length of double stranded DNA may undergo strand unwinding at numerous sites in response to promoter actions. The unraveled "bubble" of single stranded DNA has two replication forks, so assembly of new complementary strands may proceed in two directions. The polymerizations associated with several such bubbles fuse together to achieve full replication of the entire DNA double helix. Note that the events shown proceed from top to bottom in the diagram.

Repair of DNA Damage and Replication Errors

One of the benefits of the double stranded DNA structure is that it lends itself to repair, when structural damage or replication errors occur.

Several kinds of chemical change may cause damage to DNA:

- Spontaneous hydrolysis of a nucleoside removes the heterocyclic base component.
- Spontaneous hydrolysis of cytosine changes it to a uracil.
- Various toxic metabolites may oxidize or methylate heterocyclic base components.
- Ultraviolet light may dimerize adjacent cytosine or thymine bases.

All these transformations disrupt base pairing at the site of the change, and this produces a structural deformation in the double helix.. Inspection-repair enzymes detect such deformations, and use the undamaged nucleotide at that site as a template for replacing the damaged unit. These repairs reduce errors in DNA structure from about one in ten million to one per trillion.

RNA and Protein Synthesis

The genetic information stored in DNA molecules is used as a blueprint for making proteins. Why proteins? Because these macromolecules have diverse primary, secondary and tertiary structures that equip them to carry out the numerous functions necessary to maintain a living organism.

As noted in the protein, these functions include:

- Structural integrity (hair, horn, eye lenses etc.).
- Molecular recognition and signaling (antibodies and hormones).
- Catalysis of reactions (enzymes)..
- Molecular transport (hemoglobin transports oxygen).
- Movement (pumps and motors).

The critical importance of proteins in life processes is demonstrated by numerous genetic diseases, in which small modifications in primary structure produce debilitating and often disasterous consequences.

Such genetic diseases incluse Tay-Sachs, phenylketonuria (PKU), sickel cell anemia, achondroplasia, and Parkinson disease.

The unavoidable conclusion is that proteins are of central importance in living cells, and that proteins must therefore be continuously prepared with high structural fidelity by appropriate cellular chemistry.

Early geneticists identifed genes as hereditary units that determined the appearance and/ or function of an organism (*i.e.* its phenotype). We now define genes as sequences of DNA that occupy specific locations on a chromosome.

The original proposal that each gene controlled the formation of a single enzyme has since been modified as: one gene = one polypeptide. The intriguing question of how the information encoded in DNA is converted to the actual construction of a specific polypeptide has been the subject of numerous studies, which have created the modern field of Molecular Biology.

Central Dogma and Transcription

It is proposed that information flows from DNA to RNA in a process called transcription, and is then used to synthesize polypeptides by a process called translation. Transcription takes place in a manner similar to DNA replication.

A characteristic sequence of nucleotides marks the beginning of a gene on the DNA strand, and this region binds to a promoter protein that initiates RNA synthesis. The double stranded structure unwinds at the promoter site., and one of the strands serves as a template for

RNA formation, as depicted in the following diagram. The RNA molecule thus formed is single stranded, and serves to carry information from DNA to the protein synthesis machinery called ribosomes. These RNA molecules are therefore called messenger-RNA (mRNA).

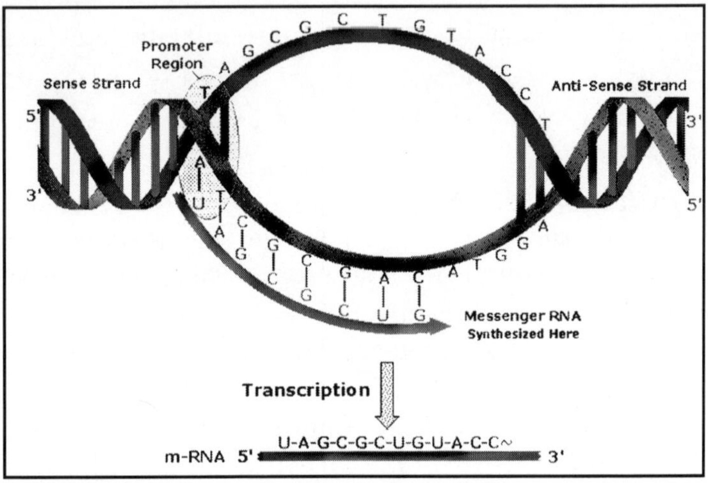

An important dstinction must be made here. One of the DNA strands in the double helix holds the genetic information used for protein synthesis. This is called the sense strand, or information strand. The complementary strand that binds to the sense strand is called the anti-sense strand and it seres as a template for generating a mRNA molecule that delivers a copy of the sense strand information to a ribosome.

The promoter protein binds to a specific nuceotide sequence that identifies the sense strand, relative to the anti-sense strand. RNA synthesis is then initiated in the 3' direction, as nucleotide triphosphates bind to complementary bases on the template strand, and are joined by phosphate diester linkages. An animation of this process for DNA replication was presented earlier. A characteristic "stop sequence" of nucleotides terminates the RNA synthesis. The messenger molecule (coloured orange above) is released into the cytoplasm to find a ribosome, and the DNA then rewinds to its double helix structure.

In eucaryotic cells the initially transcribed m-RNA molecule is usually modified and shortened by an "editing" process that removes irrelevent material. The DNA of such organisms is often thousands of times larger and more complex than that composing the single chromosome of a procaryotic bacterial cell. This difference is due in part to repetitive nucleotide sequences (ca. 25% iu the human genome). Furthermore, over 95% of human DNA is found in intervening sequences that separate genes and parts of genes. The informational DNA segments that make up genes are called exons, and the noncoding segments are called introns.

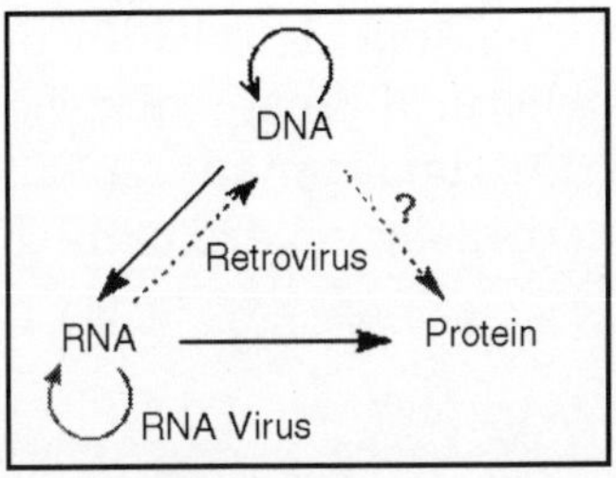

Before the mRNA molecule leaves the nucleus, the nonsense bases that make up the introns are cut out, and the informationally useful exons are joined together in a step known as RNA splicing. In this fashion shorter mRNA molecules carrying the blueprint for a specific protein are sent on their way to the ribosome factories.

The Central Dogma of molecular biology, which at first was formulated as a simple linear progression of information from DNA to RNA to Protein, is summarized in the following illustration. The replication process on the left consists of passing information from a parent DNA molecule to daughter molecules. The middle transcription process copies this information to a mRNA molecule. Finally, this information is used by the chemical machinery of the ribosome to make polypeptides.

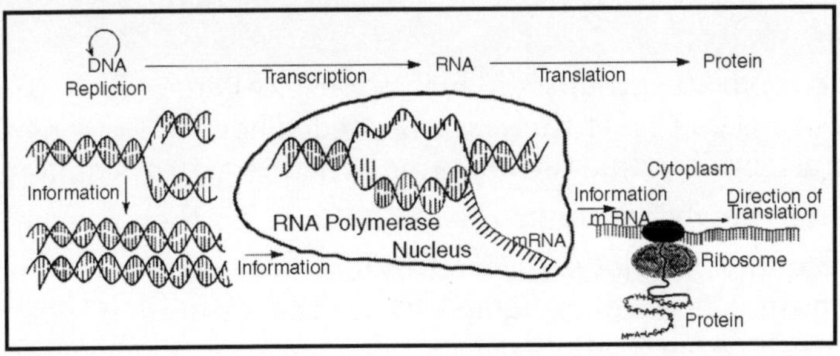

As more has been learned about these relationships, the central dogma has been refined to the representation displayed on the right. The dark blue arrows show the general, well demonstrated, information transfers noted above. It is now known that an RNA-dependent DNA polymerase enzyme, known as a reverse transcriptase, is able to transcribe a single-stranded RNA sequence into double-stranded DNA (magenta arrow).

Such enzymes are found in all cells and are an essential component of retroviruses (*e.g.* HIV), which require RNA replication of their genomes (green arrow). Direct translation of DNA information into protein synthesis (orange arrow) has not yet been observed in a living organism. Finally, proteins appear to be an informational dead end, and do not provide a structural blueprint for either RNA or DNA. In the following section the last fundamental relationship, that of structural information translation from mRNA to protein, will be described

Translation

Translation is a more complex process than transcription. This would, of course, be expected. After all, the coded messages produced by the Enigma machine could be copied easily, but required a considerable decoding effort before they could be read with understanding.

In a similar sense, DNA replication is simply a complementary base pairing exercise, but the translation of the four letter (bases) alphabet code of RNA to the twenty letter (amino acids) alphabet of protein literature is far from trivial. Clearly, there could not be a direct one-to-one correlation of bases to amino acids, so the nucleotide letters must form short words or codons that define specific amino acids.

Many questions pertaining to this genetic code were posed:

- How many RNA nucleotide bases designate a specific amino acid?

  If separate groups of nucleotides, called codons, serve this purpose, at least three are needed. There are $4^3 = 64$ different nucleotide triplets, compared with $4^2 = 16$ possible pairs.

- Are the codons linked separately or do they overlap?

  Sequentially joined triplet codons will result in a nucleotide chain three times longer than the protein it describes. If overlapping codons are used then fewer total nucleotides would be required.

- If triplet segments of mRNA designate specific amino acids in the protein, how are the codons identified?

  For the sequence ~CUAGGU~ are the codons CUA & GGU or ~C, UAG & GU~ or ~CU, AGG & U~?

- Are all the codon words the same size?

  In Morse code the most widely used letters are shorter than less common letters. Perhaps nature employs a similar scheme.

Physicists and mathematicians, as well as chemists and microbiologists all contributed to unravelling the genetic code. Although earlier proposals assumed efficient relationships that correlated the nucleotide codons uniquely with the twenty fundamental amino acids, it is now apparent that there is considerable redundancy in the code as it now operates. Furthermore, the code consists exclusively of non-overlapping triplet codons. Clever experiments provided some of the earliest breaks in deciphering the genetic code. It is found that RNA from many different organisms could initiate specific protein synthesis when combined with broken E.coli cells (the enzymes remain active). A synthetic polyuridine RNA induced synthesis of poly-phenylalanine, so the UUU codon designated phenylalanine. Likewise an alternating ~CACA~ RNA led to synthesis of a ~His-Thr-His-Thr~ polypeptide. The following table presents the resent da inter retation of the enetic code.

		Second Position					
		U	C	A	G		
First Position	U	UUU Phe [F]	UCU Ser [S]	UAU Tyr [Y]	UGU Cys [C]	U	Third Position
		UUC Phe [F]	UCC Ser [S]	UAC Tyr [Y]	UGC Cys [C]	C	
		UUA Leu [L]	UCA Ser [S]	UAA Stop	UGA Stop	A	
		UUG Leu [L]	UCG Ser [S]	UAG Stop	UGG Trp [W]	G	
	C	CUU Leu [L]	CCU Pro [P]	CAU His [H]	CGU Arg [R]	U	
		CUC Leu [L]	CCC Pro [P]	CAC His [H]	CGC Arg [R]	C	
		CUA Leu [L]	CCA Pro [P]	CAA Gln [Q]	CGA Arg [R]	A	
		CUG Leu [L]	CCG Pro [P]	CAG Gln [Q]	CGG Arg [R]	G	
	A	AUU Ile [I]	ACU Thr [T]	AAU Asn [N]	AGU Ser [S]	U	
		AUC Ile [I]	ACC Thr [T]	AAC Asn [N]	AGC Ser [S]	C	
		AUA Ile [I]	ACA Thr [T]	AAA Lys [K]	AGA Arg [R]	A	
		AUG Met [M]	ACG Thr [T]	AAG Lys [K]	AGG Arg [R]	G	
	G	GUU Val [V]	GCU Ala [A]	GAU Asp [D]	GGU Gly [G]	U	
		GUC Val [V]	GCC Ala [A]	GAC Asp [D]	GGC Gly [G]	C	
		GUA Val [V]	GCA Ala [A]	GAA Glu [E]	GGA Gly [G]	A	
		GUG Val [V]	GCG Ala [A]	GAG Glu [E]	GGG Gly [G]	G	

Note that this is the RNA alphabet, and an equivalent DNA codon table would have all the U nucleotides replaced by T. Methionine and tryptophan are uniquely represented by a single codon. At the other extreme, leucine is represented by eight codons. The average redundancy for the twenty amino acids is about three. Also, there are three stop codons that terminate polypeptide synthesis.

RNA Codons for Protein Synthesis

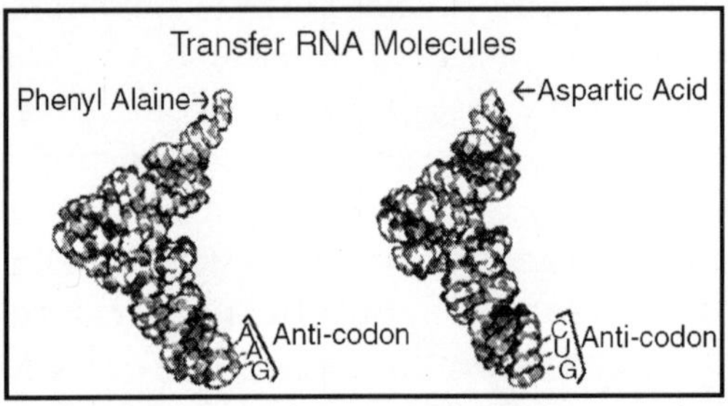

The translation process is fundamentally straightforward. The mRNA strand bearing the transcribed code for synthesis of a protein interacts with relatively small RNA molecules (about 70-nucleotides) to which individual amino acids have been attached by an ester bond at the 3'-end. These transfer RNA's(tRNA) have distinctive three-dimensional structures cosisting of loops of single-stranded RNA connected by double stranded segments. This cloerleaf secondary structure is further wrapped into an "L-shaped" assembly, having the amino acid at the end of one arm, and a characteristic anti-codon region at the other end. The anti-codon consists of a nucleotide triplet that is the complement of the amino acid's codon(s). Models of two such tRNA molecules are shown to the right. When read from the top to the bottom, the anti-codons depicted here should complement a codon in the previous table. A cell's protein synthesis takes place in organelles called ribosomes. Ribosomes are complex structures made up of two distinct and separable subunits (one about twice the size of the other). Each subunit is composed of one or two RNA molecules (60-70%) associated with 20 to 40 small proteins (30-40%). The ribosome accepts a mRNA molecule, binding initially to a characteristic nucleotide sequence at the 5'-end (coloured light blue in the following diagram).

This unique binding assures that polypeptide synthesis starts at the right codon. AtRNA molecule with the appropriate ant-codon thn attaches at the starting point and this is followed by a series of adjacent tRNA attachments, peptide bond formation and shifts of the ribosome along the mRNA chain to expose new codons to the ribosomal chemistry.

The following diagram is designed as a slide show illustrating these steps. The outcome is snthesis of a polypeptide chain corresponding to the mRNA blueprint. A "stop codon" at a designated position on the mRNA terminates the synthesis by introduction of a "Release Factor".

Drug acting on DNA

Mechanism of Intercalating

Intercalating agents wedge between bases along the DNA. The intercalated drug molecules affect the structure of the DNA, preventing polymerases and other DNA binding proteins from functioning properly. The result is prevention of DNA synthesis, inhibition of transcription and induction of mutations. The effect of principal alkaloids (sanguinarine, chelerythrine, coptisine, chelidonine) of greater celandine Chelidonium majus L., as well as the alkaloids from Colchicum autumnale L.(colchicine and colchamine) on calcium accumulation and oxidative phosphorylation in rat liver mitochondria has been studied.

The obtained data were compared with DNA intercalating properties of alkaloids detected by the method of thermodenaturation (DNA melting curve plots). It was found that chelerythrin and sanguinarine blocked absorption and accumulation of calcium cations and inhibited oxidative phosphorylation, while the coptisine significantly diminished those indices. Chelidonine, colchicines and colchamine had no influence on the studied characteristics. The effect of alkaloids upon mitochondria functional state correlated tightly with their DNA intercalating properties: chelerythrine and sanguinarine were strong intercalators, while coptisine was a weak one, and chelidonine, colchicine and colchamine did not interact with DNA and caused no changes in its melting point.

Correlation coefficient between the intercalating properties of alkaloids and their inhibition of calcium accumulation was.89, and with their oxidative phosphorylation inhibition -.93. It is suggested that the effect of studied alkaloids upon functional properties of mitochondria can be mediated by mtDNA. DNA topoisomerase I is essential for cellular metabolism and survival. It is also the target of a novel class of anticancer drugs active against previously refractory solid tumors, the camptothecins. The present review describes the topoisomerase I catalytic mechanisms with particular emphasis on the cleavage complex that represents the enzyme's catalytic intermediate and the site of action for camptothecins. Roles of topoisomerase I in DNA replication, transcription and recombination are also reviewed. Because of the importance of topoisomerase I as a chemotherapeutic target, we review the mechanisms of action of camptothecins and the other topoisomerase I inhibitors identified to date.

The DNA intercalating agents 4'-(9-acridinyl-amino) methanesulphon-m-anisidide (m-AMSA) and adriamycin were studied by using filter elution methods to measure DNA single-strand breaks (SSB's), DNA-protein cross-links (DPC's), and double-stranded breaks (DSB's) in mouse leukemia L1210 cells. Both compounds produced SSB's and DPC's at nearly 1:1 ratios. The SSB's and DPC's were shown to be localized with respect to each other; this was inferred from the finding that filter assays based on protein adsorption completely prevented the elution of the DNA single-strand segments between SSB's. In the case of m-AMSA, which produces relatively high frequencies of DNA lesions, the possibility that a protein bridges across the SSB was excluded by alkaline sedimentation studies.

The o-AMSA isomer is much less cytotoxic than m-AMSA and did not produce protein-associated strand breaks. The simplest model to explain the results is that a protein becomes covalently bound to either the 3' or the 5' termini of the intercalator-induced strand breaks.

At moderately cytotoxic doses, m-AMSA yielded much larger frequencies of protein-associated SSB's than did adriamycin. m-AMSA-induced protein-associated SSB's saturated at approximately 60000 per cell over a concentration range in which m-AMSA uptake by the cells was proportional to the drug concentration. m-AMSA was found to enter and exit from cells very rapidly at 37 degrees C; protein-associated SSB's and DSB's also appeared and disappeared rapidly. At reduced temperature, however, the appearance and disappearance of protein-associated SSB's could be blocked while m-AMSA entry and exit still occurred.

The saturation behaviour and temperature dependence suggest that the formation and disappearance of protein-associated strand breaks is enzymatic. The simplest hypothesis is that the linked protein is a nuclease, such as a topoisomerase, which becomes bound to one terminus of the strand break it produces. It is proposed that topoisomerases producing SSB's and DSB's are stimulated to different degrees by different intercalators. Topoisomerase II mediated DNA scission induced by both a nonintercalating agnt [4'-demethylepipodophyllotoxin 4-(4,6-O-ethylidene-beta-D-glucopyranoside)(VP-16)] and an intercalator [4'-(9-acridinylamino) methanesulphon-m-anisidide (m-AMSA)] wasstudied as a function of proliferation in Chinese hamster ovary (CHO), HeLa, an mouse leukemia L1210 cell lines. Log-phase CHO cells exhibited dose-dependent drug-induced DNA breaks, while plateau cells were found to be resistant to the effects of VP-16 and m-AMSA. Neither decreased viability nor altered drug uptake accounted for the drug resistance of these confluent cells. In contrast to CHO cells, plateau-phase HeLa and L1210 cells remained sensitive to VP-16 and m-AMSA. Recovery of drug sensitivity by plateau-phase CHO cells was found to reach a maximum approximately 18 h after these cells regained exponential growth and was independent of DNA synthesis.

DNA strand break frequency correlated with cytotoxicity in CHO cells; log cells demonstrated an inverse log linear relationship between drug dose (or DNA damage) and colony survival, whereas plateau-derived colony survival was virtually unaffected by increasing drug dose. Topoisomerase II activity, whether determined by decatenation of kinetoplast DNA, by cleavage of pBR322 DNA, or by precipitation of the DNA-topoisomerase II complex, was uniformly severalfold greater in log-phase CHO cells compared to plateau-phase cells.

The biochemical characteristics of the formation and disappearance of intercalator-induced DNA double-strand breaks (DSB) were studied in nuclei from mouse leukemia L1210 cells by using filter elution methodology [Bradley, M. O.,& Kohn, K.W.(1979) Nucleic Acids Res. 7, 793-804]. The three intercalators used were 4'-(9-acridinylamino)-methanesulphon-m-anisidide (m-AMSA), 5-iminodaunorubicin (5-ID), and ellipticine. These compounds differ in that they produced predominantly DNA single-strand breaks (SSB)(m-AMSA) or predominantly DNA double-strand breaks (ellipticine) or a mixture of both SSB and DSB (5-ID) in whole cells. In isolated nuclei, each intercalator produced DSB at a frequency comparable to that which is produced in whole cells. Moreover, these DNA breaks reversed within 30 min after drug removal. It thus appeared that neither ATP nor other nucleotides were necessary for intercalator-dependent DNA nicking-closing reactions. The formation of the intercalator-induced DSB was reduced at ice temperature. Break formation was also reduced in the absence of magnesium, at a pH above 6.4 and at NaCl concentrations above 200 mM. In the presence of ATP and AP analogues, the intercalator-induced cleavage was enhanced. These results suggest that the intercalator-induced DSB are enzymatically mediated and that the enzymes involved in these reactions can catalyze DNA double-strand cleavage and rejoining in the absence of ATP, although the occupancy of an ATP binding site might convert the enzyme to a form more reactive to intercalators. Three inhibitors of DNA topoisomerase II—novobiocin, nalidixic acid, and norfloxacin—reduced the formation of DNA strand breaks.

Mechanisms of Alkylating Aents

Alkylating agents work by three different mechanisms all of which achieve the same end result - disruption of DNA function and cell death.In the first mechanism an alkylating agent (represented in the figure below as a pink star) attaches alkyl groups (small carbon compounds-depicted as pink triangles) to DNA bases. This alteration results in the DNA being fragmented by repair enzymes in their attempts to replace the alkylated bases (frame 3 of the diagram below). Alkylated bases prevent DNA synthesis and RN transcription from the affected DNA.

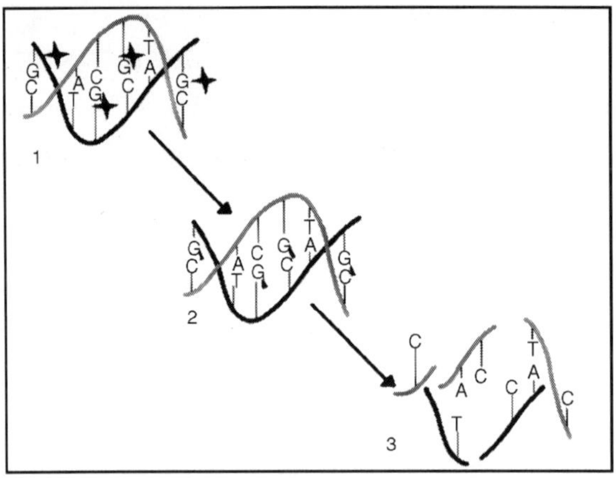

A second mechanism by which alkylating agents cause DNA damage is the formation of cross-bridges, bonds between atoms in the DNA (pink linkages below). In this process, two bases are linked together by an alkylating agent that has two DNA binding sites. Bridges can be formed within a single molecule of DNA (as shown below) or a cross-bridge may connect two different DNA molecules. Cross-linking prevents DNA from being separated for synthesis or transcription.

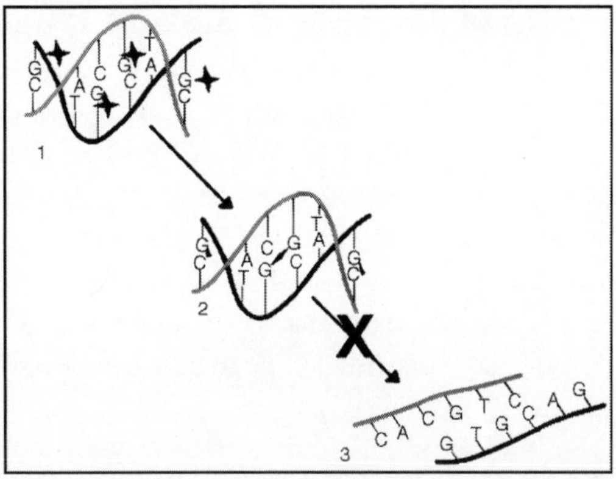

The third mechanism of action of alkylating agents is the induction of mispairing of the nucleotidesleading to mutations. In a normal DNA double helix, A always pairs with (is across from) T and G always pairs with C. As the figure below shows, alkylated G bases may erroneously pair with Ts. If this altered pairing is not corrected it may lead to a permanent mutation.

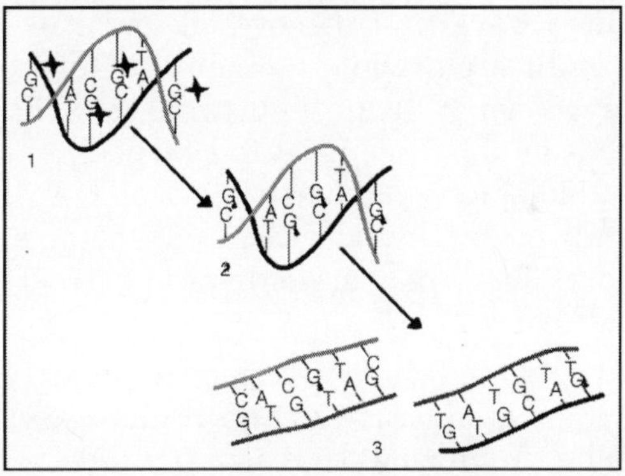

Alkylating agents are one of the earliest and most commonly used chemotherapy agents used for cancer treatments. Their use in cancer treatments started in early 1940s. Majority of alkaline agents are active or dormant nitrogen mustards, which are poisonous compound initially used for certain military purposes. Chlorambucil, Cyclophosphamide, CCNU, Melphalan, Procarbazine, Thiotepa, BCNU, and Busulphan are some of the commonly used alkylating agents.

Although they might differ in their clinical activity, action mechanism of all alkylating agents is the same. These agents work directly on the DNA and prevent the cell division process by cross-linking and breaking the DNA strands and causing abnormal base pairing. When a DNA is altered in this manner, undesired cellular activity comes to a halt and the cell dies eventually. Alkylating chemotherapy drugs are effective during all phases of cell cycle. Therefore, they are used to treat a larg number of cancers. However, they are more effective in treating slow-growing cancers such as solid tumors and leukemia. Long term use of alkylating agents can lead to permanent infertility by decreasing sperm production in males, and causing menstruation cessation in females. Many alkylating agents can also lead to secondary cancers such as Acute Myeloid Leukemia, years after the therapy.

Antimetabolites

Structure of antimetabolites (antineoplastic agents) is similar to certain compounds such as vitamins, amino acids, and precurors of DNA or RNA, found naturally in human body. Antimetabolites help in treatment cancer by inhibiting cell division thereby hindering the growth of tumor cells. These agents get incorporated in the DNA or RNA to interfere with the process of division of cancer cells.

Antimetabolites were first discovered and it is found that folic acid analog can reduce childhood leukemia. Out of 16 patients he tested, 10 displayed hematologic improvement. This discovery laid the foundation that enabled scientist to synthesize many new agents that could inhibit biological enzymatic reactions. Antimetabolites are found to be useful in treating chronic and acute cases of leukemia and various tumors. They are commonly used to treat gastrointestinal tract, breast, and ovary tumors.

Methotraxate, which is a commonly used antimetabolites chemotherapy agent, is effective in the S-phase of the cell cycle. It works by inhibiting an enzyme that is essential for DNA synthesis. 6-mercaptopurine and 5-fluorouracil (5FU) are two other commonly used antimetabolites. 5-Fluorouracil (5-FU) works by interfering with the DNA components, nucleotide, to stop DNA synthesis. This drug is used to treat many different types of cancers including breast, esophageal, head, neck, and gastric cancers. 6-mercaptopurine is an analogue of hypoxanthine and is commonly used to treat Acute Lymphoblastic Leukemia (ALL). Other popular antimetabolite chemotherapy drugs are Thioguanine, Cytarabine, Cladribine. Gemcitabine, and Fludarabine.

Anthracyclines

Anthracyclines were developed between 1970s and 1990s and are daunosamine and tetra-hydronaphthacenedione-based chemotherapy agents. These compounds are cell-cycle nonspecific and are used to treat a large number of cancers including lymphomas, leukemia, and uterine, ovarian, lung and breast cancers.

Anthracyclines drugs are developed from natural resources. For instance, daunorubicin is developed by isolating it from soil-dwelling fungus Streptomyces. Similarly, Doxorubicin, which is another commonly used anthracycline chemotherapy agent, is isolated from mutated strain of Streptomyces. Although both the drugs have similar clinical action mechanisms, doxorubicin is more effective in treating solid tumors. Idarubicin, Epirubicin, and Mitoxantrone are few of the other commonly used anthracycline chemotherapy drugs. Anthracyclines work by forming free oxygen radicals that breaks DNA strands thereby inhibiting DNA synthesis and function. These chemotherapeutic agents form a complex with DNA and enzyme to inhibit the topoisomerase enzyme. Topoisomerase is an enzyme class that causes the supercoiling of DNA, allowing DNA repair, transcription, and replication.

One of the main side effects of anthracyclines is that it can damage cells of heart muscle along with the DNA of cancer cell leading to cardiac toxicity.

Antitumor Antibiotics

Antitumor antibiotics are also developed from the soil fungus Streptomyces. These drugs are widely used to treat and suppress development of tumors in the body. Similar to anthracyclines, antitumor antibiotics drugs also form free oxygen radicals that result in DNA strand breaks, killing the growth of cancer cells. In most of the cases, these drugs are used in combination with other chemotherapy agents. Bleomycin is one of the commonly used antitumor antibiotic used to treat testicular cancer and hodgkin's lymphoma. The most serious side effect of this drug is lung toxicity that occurs when the oxygen radical formed by the antitumor antibiotics damages lung cells along with the cancer cells.

Monoclonal Antibodies

Monoclonal antibodies are one of the newer chemotherapy agents approved for cancer treatment by the Food and Drug Administration (FDA) in 1997. Alemtuzumab (Campath), Bevacizumab (Avastin), Cetuximab (Erbitux), Gemtuzumab (Mylotarg), Ibritumomab (Zevalin), Panitumumab (Vectibix), Rituximab (Rituxan), Tositumomab (Bexxar), and Trastuzumab (Herceptin) are some of the FDA approved monoclonal drugs used in chemotherapeutic cancer treatments. The treatment is known to be useful in treating colon, lung, head, neck, and breast cancers. Some of the monoclonal drugs are used to treat chronic lymphocytic leukemia, acute myelogenous leukemia, and non-Hodgkin's lymphoma.

Monoclonal antibodies work by attaching to certain parts of the tumor-specific antigens and make them easily recognizable by the host's immune system. They also prevent growth of cancer cells by blocking the cell receptors to which chemicals called 'growth factors' attach promoting cell growth. Monoclonal antibodies can be combined with radioactive particles and other powerful anticancer drugs to deliver them directly to cancer cells. Using this method, long term radioactive treatment and anticancer drugs can be given to patients without causing any serious harm to other healthy cells of the body.

Platinum

Platinum-based natural metal derivatives were found to be useful for cancer treatments around 150 years ago with the synthesis of cisplatin. However, there clinical use did not commence until 30 years ago. Platinum-based chemotherapy agents work by cross-linking subunits of DNA. These agents act during any part of cell cycle and help in treating cancer by impairing DNA synthesis, transcription, and function.

Cisplatin, although found to be useful in treating testicular and lung cancer, is highly toxic and can severely damage the kidneys of the patient. Second generation platinum-complex carboplatin is found to be much less toxic in comparison to cisplatin and has fewer kidney-related side effects. Oxaliplatin, which is third generation platinum-based complex, is found to be helpful in treating colon cancer. Although, oxaliplatin does not cause any toxicity in kidney it can lead to severe neuropathies.

Plant Alkaloid

Plant alkaloid chemotherapy agents, as the name suggests, are plant derivatives. They are cell-specific chemotherapy agents. However, the cycle affected is based on the drug used for the treatment. They are primarily categorized into four groups: topoisomerase inhibitors, vinca alkaloids, taxanes, and epipodophyllotoxins. Plant alkaloids are cell-cycle specific, but the cycle affected varies from drug to drug. Vincristine (Oncovin) is a plant alkaloid of interest in mesothelioma treatment.

Topoisomerase Inhibitors

Topoisomerase inhibitors are chemotherapy agents are categorized into Type I and Type II Topoisomerases inhibitors and they work by interfering with DNA transcription, replication, and function to prevent DNA supercoiling.

- *Type I Topoisomerase inhibitors:* These chemotherapy agents are extracted from the bark and wood of the Chinese tree Camptotheca accuminata. They work by forming a complex with topoisomerase DNA. This in turn suppresses the function of topoisomerase.

Camptothecins which includes irinotecan and topotecan are commonly used type I topoisomerase inhibitors, first discovered in the late 1950s.

- *Type II Topoisomerase inhibitors:* These are extracted from the alkaloids found in the roots of May Apple plants. They work in the in the work in the late S and G2 phases of the cell cycle.

Amsacrine, etoposide, etoposide phosphate, and teniposide are some of the examples of type II topoisomerase inhibitors.

Vinca Alkaloids

Vinca alkaloids are derived from the periwinkle plant, Vinca rosea (Catharanthus roseus) and are known to be used by the natives of Madagascar to treat diabetes.

Although not useful in controlling diabetes, vinca alkaloids, are useful in treating leukemias. They are effective in the M phase of the cell cycle and work by inhibiting tubulin assembly in microtubules. Vincristine, Vinblastine, Vinorelbine, and Vindesine are some of the popularly used vinca alkaloid chemotherapy agents used today. Major side effect of vinca alkaloids is that they can cause neurotoxicity in patients.

Taxanes

Taxanes are plant alkaloids that were first developed in 1963 by isolating it from first isolated from the bark of the Pacific yew tree, Taxus brevifolia in 1963. Paclitaxel, which is the active components of taxanes was first discovered in 1971 and was made available for clinical use in the year 1993. Taxanes also work in the M-phase of the cell cycle and inhibit the function of microtubules by binding with them. Paclitaxel and docetaxel are commonly used taxanes. Taxanes chemotherapy agents are used to treat a large array of cancers including breast, ovarian, lung, head and neck, gastric, esophageal, prostrate and gastric cancers. The main side effect of taxanes is that they lower the blood counts in patients.

Anti-Cancer Drug

Introduction

The available anticancer drugs have distinct mechanisms of action which may vary in their effects on different types of normal and cancer cells. A single "cure" for cancer has proved elusive since there is not a single type of cancer but as many as 100 different types of cancer. In addition, there are very few demonstrable biochemical differences between cancerous cells and normal cells. For this reason the effectiveness of many anticancer drugs is limited by their toxicity to normal rapidly growing cells in the intestinal and bone marrow areas. A final problem is that cancerous cells which are initially suppressed by a specific drug may develop a resistance to that drug. For this reason cancer chemotherapy may consist of using several drugs in combination for varying lengths of time.

Cancer Chemotherapy

Chemotherapy drugs, are sometimes feared because of a patient's concern about toxic effects. Their role is to slow and hopefully halt the growth and spread of a cancer. There are three goals associated with the use of the most commonly-used anticancer agents.

- Damage the DNA of the affected cancer cells.
- Inhibit the synthesis of new DNA strands to stop the cell from replicating, because the replication of the cell is what allows the tumor to grow.
- Stop mitosis or the actual splitting of the original cell into two new cells. Stopping mitosis stops cell division (replication) of the cancer and may ultimately halt the progression of the cancer.

Unfortunately, the majority of drugs currently on the market are not specific, which leads to the many common side effects associated with cancer chemotherapy. Because the common approach of all chemotherapy is to decrease the growth rate (cell division) of the cancer cells, the side effects are seen in bodily systems that naturally have a rapid turnover of cells iincluding skin, hair, gastrointestinal, and bone marrow. These healthy, normal cells, also end up damaged by the chemotherapy programme.

Catagories of Chemotherapy Drugs

In general, chemotherapy agents can be divided into three main categories based on their mechanism of action.

Stop the synthesis of pre DNA molecule building blocks: These agents work in a number of different ways. DNA building blocks are folic acid, heterocyclic bases, and nucleotides, which are made naturally within cells. All of these agents work to block some step in the formation of nucleotides or deoxyribon-ucleotides (necessary for making DNA). When these steps are blocked, the nucleotides, which arethe building blocks of DNA and RNA, can not be synthesized. Thus the cells can not replicate because they can nnot make DNA without the nucleotides.

Examples of drugs in this class include:

- Methotrexate (Abitrexate®),
- Fluorouracil (Adrucil®),
- Hydroxyurea (Hydrea®), and
- Mercaptopurine (Purinethol®).

Directly damage the DNA in the nucleus of the cell:

These agents chemically damage DNA and RNA. They disrupt replication of the DNA and either totally halt replication or cause the manufacture of nonsense DNA or RNA (*i.e.* the new DNA or RNA does not code for anything useful).

Examples of drugs in this class include cisplatin (Platinol®) and antibiotics - daunorubicin (Cerubidine®), doxorubicin (Adriamycin®), and etoposide (VePesid®).

Effect the synthesis or breakdown of the mitotic spindles:

Mitotic spindles serve as molecular railroads with "North and South Poles" in the cell when a cell starts to divide itself into two new cells. These spindles are very important because they help to split the newly copied DNA such that a copy goes to each of the two new cells during cell division. These drugs disrupt the formation of these spindles and therefore interrupt cell division.

Examples of drugs in this class of miotic disrupters include: Vinblastine (Velban), Vincristine (Oncovin) and Pacitaxel (Taxol).

Methotrexate: Methotrexate inhibits folic acid reductase which is responsible for the conversion of folic acid to tetrahydrofolic acid. At two stages in the biosynthesis of purines (adenine and guanine) and at one stage in the synthesis of pyrimidines (thymine, cytosine, and uracil), one-carbon transfer reactions occur which require specific coenzymes synthesized in the cell from tetrahydrofolic acid. Tetrahydrofolic acid itself is synthesized in the cell from folic acid with the help of an enzyme, folic acid reductase. Methotrexate looks a lot like folic acid to the enzyme, so it binds to it thinking that it is folic acid. In fact, methotrexate looks so good to the enzyme that it binds to it quite strongly and inhibits the enzyme. Thus, DNA synthesis cannot proceed because the coenzymes needed for one-carbon transfer reactions are not produced from tetrahydrofolic acid because there is no tetrahydrofolic acid. Again, without DNA, no cell division.

5-Fluorouracil: 5-Fluorouracil (5-FU; Adrucil, Fluorouracil, Efudex, Fluoroplex) is an effective pyrimidine antimetabolite. Fluorouracil is synthesized into the nucleotide, 5-fluoro-2-deoxyuridine. This product acts as an antimetabolite by inhibiting the synthesis of 2-deoxythymidine because the carbon - fluorine bond is extremely stable and prevents the addition of a methyl group in the 5-position. The failure to synthesize the thymidine nucleotide results in little or no production of DNA.

Two other similar drugs include: gemcitabine (Gemzar) and arabinosylcytosine (araC). They all work through similar mechanisms.

Hydroxyurea: Hydroxyurea blocks an enzyme which converts the cytosine nucleotide into the deoxy derivative. In addition, DNA synthesis is further inhibited because hydroxyurea blocks the incorporation of the thymidine nucleotide into the DNA strand.

Mercaptopurine: Mercaptopurine, a chemical analog of the purine adenine, inhibits the biosynthesis of adenine nucleotides by acting as an antimetabolite. In the body, 6-MP is converted to the corresponding ribonucleotide. 6-MP ribonucleotide is a potent inhibitor of the conversion of a compound called inosinic acid to adenine Without adenine, DNA cannot be synthesized. 6-MP also works by being incorporated into nucleic acids as thioguanosine, rendering the resulting nucleic acids (DNA, RNA) unable to direct proper protein synthesis. Adenine and Mercaptopurine - Chime in new window

Thioguanine: Thioguanine is an antimetabolite in the synthesis of guanine nucleotides.

Alkylating Agents: Alkylating agents involve reactions with guanine in DNA. These drugs add methyl or other alkyl groups onto molecules where they do not belong. This in turn inhibits their correct utilization by base pairing and causes a miscoding of DNA.

In the first mechanism an alkylating agent attaches alkyl groups to DNA bases. This alteration results in the DNA being fragmented by repair enzymes in their attempts to replace the alkylated bases. A second mechanism by which alkylating agents cause DNA damage is the formation of cross-bridges, bonds between atoms in the DNA. In this process, two bases are linked together by an alkylating agent that has two DNA binding sites. Cross-linking prevents DNA from being separated for synthesis or transcription. The third mechanism of action of alkylating agents causes the mispairing of the nucleotides leading to mutations.

There are six groups of alkylating agents: nitrogen mustards; ethylenimes; alkylsulphonates; triazenes; piperazines; and nitrosureas. Cyclosporamide is a classical example of the role of the host metabolism in the activation of an alkylating agent and is one or the most widely used agents of this class. It was hoped that the cancer cells might posses enzymes capable of accomplishing the cleavage, thus resulting in the selective production of an activated nitrogen mustard in the malignant cells. Compare the top and bottom structures in the graphic on the left.

Antibiotics: A number of antibiotics such as anthracyclines, dactinomycin, bleomycin, adriamycin, mithramycin, bind to DNA and inactivate it. Thus the synthesis of RNA is prevented.

General properties of these drugs include: interaction with DNA in a variety of different ways including intercalation (squeezing between the base pairs), DNA strand breakage and inhibition with the enzyme topoisomerase II. Most of these compounds have been isolated from natural sources and antibiotics. However, they lack the specificity of the antimicrobial antibiotics and thus produce significant toxicity.

Th anthracyclines are among the most important antitumor drugs available. Doxorubicin is widely used for the treatment of several solid tumors while daunorubicin and idarubicin are used exclusively for the treatment of leukemia.

These agents have a number of important effects including: intercalating (squeezing between the base pairs) with DNA affecting many functions of the DNA including DNA and RNA synthesis. Breakage of the DNA strand can also occur by inhibition of the enzyme topoisomerase II.

Dactinomycin (Actinomycin D): At low concentrations dactinomycin inhibits DNA directed RNA synthesis and at higher concentrations DNA synthesis is also inhibited. All types of RNA are affected, but ribosomal RNA is more sensitive. Dactinomycin binds to double stranded DNA, permitting RNA chain initiation but blocking chain elongation. Binding to the DNA depends on the presence of guanine.

Mitotic Disrupters: Plant alkaloids like vincristine prevent cell division, or mitosis. There are several phases of mitosis, one of which is the metaphase. During metaphase, the cell pulls duplicated DNA chromosomes to either side of the parent cell in structures called "spindles". These spindles ensure that each new cell gets a full set of DNA. Spindles are microtubular fibres formed with the help of the protein "tubulin". Vincristine binds to tubulin, thus preventing the formation of spindles and cell division. In contrast to other microtubule antagonists, taxol disrupts the equilibrium between free tubulin and mircrotubules by shifting it in the direction of assembly, rather than disassembly. As a result, taxol treatment causes both the stabilization of microtubules and the formation of abnormal bundles of microtubules. The net effect is still the disruption of mitosis.

Ribonucleic Acid

Ribonucleic acid (RNA) is a biologically important type of molecule that consists of a long chain of nucleotide units. Each nucleotide consists of a nitrogenous base, a ribose sugar, and a phosphate. RNA is very similar to DNA, but differs in a few important structural details: in the cell, RNA is usually single-stranded, while DNA is usually double-stranded; RNA nucleotides contain ribose while DNA contains deoxyribose (a type of ribose that lacks one oxygen atom); and RNA has the base uracil rather than thymine that is present in DNA.

Ribonucleic acid (RNA) is a linear molecule composed of four types of smaller molecules called ribonucleotide bases: adenine (A), cytosine (C), guanine (G), and uracil (U). RNA is often compared to a copy from a reference book, or a template, because it carries the same information as its DNA template but is not used for long-term storage. Each ribonucleotide base consists of a ribose sugar, a phosphate group, and a nitrogenous base. Adjacent ribose nucleotide bases are chemically attached to one another in a chain via chemical bonds called phosphodiester bonds. Unlike DNA, RNA is usually single-stranded. Additionally, RNA contains ribose sugars rather than deoxyribose sugars, which makes RNA more unstable and more prone to degradation. RNA is synthesized from DNA by an enzyme known as RNA polymerase during a process called transcription. The new

RNA sequences are complementary to their DNA template, rather than being identical copies of the template. RNA is then translated into proteins by structures called ribosomes. There are three types of RNA involved in the translation process: messenger RNA (mRNA), transfer RNA (tRNA), and ribosomal RNA (rRNA). Although some RNA molecules are passive copies of DNA, many play crucial, active roles in the cell. For example, some RNA molecules are involved in switching genes on and off, and other RNA molecules make up the critical protein synthesis machinery in ribosomes.

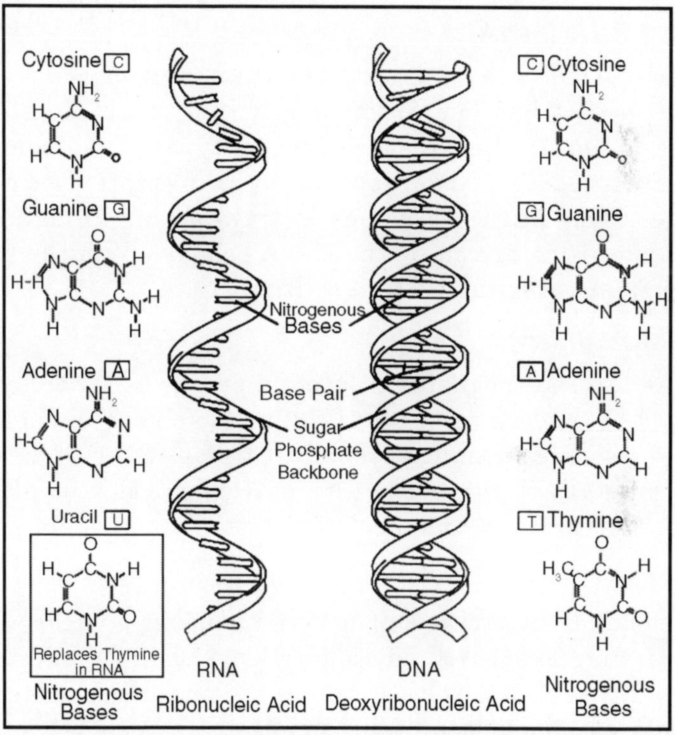

RNA is transcribed from DNA by enzymes called RNA polymerases and is generally further processed by other enzymes. RNA is central to the synthesis of proteins. Here, a type of RNA called messenger RNA carries information from DNA to structures called ribosomes.

These ribosomes are made from proteins and ribosomal RNAs, which come together to form a molecular machine that can read messenger RNAs and translate the information they carry into proteins. There are many RNAs with other roles – in particular regulating which genes are expressed, but also as the genomes of most viruses.

Messenger RNA (mRNA) and its Precursors

The primary RNA transcripts are produced by copying the DNA of a protein-encoding gene.[1] Subsequent processing steps include addition, removal, and modification of nucleotides as well as splicing events that excise internal segments. The mature mRNA is translated to produce a protein whose amino acid sequence is determined by the sequence of the coding region in the gene. The messenger RNA, as the name implies, is the molecule that carries the message from the gene to the protein synthesis machinery.

Ribosomal RNA (rRNA)

The ribosomes are the most important part of the translation machinery and it has long been known that much of the mass of ribosomes is due to the presence several types of ribosomal RNA. These are noncoding RNAs produced by transcription of ribosomal RNA genes.[3] One of the key steps in translation—formation of the peptide bond—is catalyzed by the rRNA component of the ribosome. It is the major catalytic RNA in cells.

Transfer RNA (tRNA)

tRNAs are intermediates in protein synthesis. There are many different tRNA molecules in every cell and each one binds a specific amino acid, yielding an aminoacylated-tRNA (aa-tRNA). Each different aminoacylated tRNA interacts with a particular codon in mRNA thus delivering the correct amino acid to to the site of protein synthesis.

Small RNAs

The small RNAs represent a heterogeneous category of RNAs covering a wide ranges of functions. Some of them have catalytic functions—RNAse P is the classic example. Some of them are structural components of ribonucleoprotein complexes (*e.g.* signal recognition particle). Some of them are guide RNAs involved in various processing events. The best known examples of guide RNAs are the small RNAs of the spliceosome complexes that mediate the splicing of mRNA precursors.[2]Other small RNAs were known to be involved in the regulation of gene expression.

Genomic RNA

Some viruses, notably retroviruses, have an RNA genome instead of a DNA genome. In addition, the mobility of various transposons is due to an intermediate RNA copy of the transposon sequence (retrotransposons). This was the state of knowledge 25 years ago. Since then, the study of RNA has made remarkable progress. Our knowledge of all the fundamental processes—transcription, processing, and catalysis—has expanded enormously.

The biggest change is in the area of small RNAs. Today there are several categories of small RNAs—siRNA, microRNA, piRNA—that were only discovered in the past 10-15 years. The functions of these small RNA molecules are still being worked out. There's little doubt that some of them have important biological roles but there's considerable controversy over what percentage might be artifacts of one sort or another.

RNA replaces DNA as the genetic material. As with DNA, RNA follows specific base pairing rules, except that in RNA the base uracil replaces the base thymine (*i.e.*, instead of an adenine-thymine or A-T pairing, there is an adenine-uracil or A-U pairing). Accordingly, when RNA acts as a carrier of genetic information, uracil replaces thymine in the genetic code.

In humans, messenger RNA (mRNA) is the product of transcription and acts to convey genetic information from the nucleus to the protein assembly complex at the ribosome. The ribsome is composed of ribosomal RNA (rRNA) and other proteins. Transfer RNAs (tRNA) act to catalyze the translation process by acting as carriers of specific amino acids. Because tRNAs bind to specific sites on the strand of mRNA, the sequence of amino acids subsequently inserted into the synthesized protein is both specific and genetically determined by the nucleotide sequence in DNA from which the mRNA strand was originally transcribed.

Other forms of RNA perform important roles in other biochemical reactions. Regardless of function, RNA is a biopolymer made up of ribonucleotide units and is present in all living cells and some viruses. The chemical units of RNA are ribonucleotide monomers consisting of a ribose sugar ($C_5H_{10}O_5$) phosphorylated at the third carbon (C3) and linked to one of four bases through a type of chemical linkage formed between a sugar and a base by a condensation reaction (glycosidic bond). The four bases found in RNA are adenine (A), guanine (G), cytosine (C), and uracil (U). Other bases may also be found, although they are generally modified versions of these four (*e.g.*, methylated bases are found in parts of tRNA).

The single nucleotides (monomers) of RNA form a linear chain by linking their phosphate groups and sugars in phosphodiester bonds.

RNA does not form a double stranded alpha-helix as does DNA. In some parts of the RNA molecule, there is folding into alpha-helical-like regions. Corresponding to their unique functions, messenger RNA (mRNA), ribosomal RNA (rRNA), and transfer RNA (tRNA) all have different three-dimensional structures. In higher eukaryotic organisms, different RNAs are found distributed throughout the cell—in the nucleus, cytoplasm, and also in cytoplasmic organelles such as mitochondria and, in plants, chloroplasts.

The nucleus is the chief site of RNA synthesis and the source of all cytoplasmic RNA, while mitochondria and chloroplasts synthesize their RNA from their own DNA. rRNA is synthesized by the nucleoli within the nucleus, while the high molecular weight precursor to cytoplasmic mRNA, sometimes termed heterogeneous nuclear or hnRNA, is transcribed on the DNA chromatin. Low molecular weight RNA also occurs in the nucleus and consists partly of tRNA and partly of RNA, which has a regulatory function in geneactivation. The cytoplasm contains tRNA and rRNA in the ribosomes and mRNA in polysomes, or polyribosomes. The latter are the structural units of protein biosynthesis, consisting of several ribosomes attached to a strand of mRNA.

The function of mRNA is to transcribe the information held in DNA. In the cells of eukaryotic organisms, the first transcriptional product is the long, heterogenous nuclear RNA, or hnRNA. This contains both the nucleotide sequences eventually transcribed into polypeptides and large tracts of sequences not translated. Non-translated sequences are termed introns (or intervening sequences). Removal of introns, and other untranslated portions of the molecule, edits hnRNA into mRNA molecules. After editing removes as much as 90% of hnRNA, the resulting mRNA molecules are transported into the cytoplasm.

rRNA is located within ribosomes, the sites of protein biosynthesis. Ribosomes are large ellipsoid cytoplasmic organelles consisting of RNA and protein.

tRNA, the smallest known functional RNA, is essential for protein biosynthesis. Its purpose is to transfer a specific amino acid from the cytoplasm and incorporate it into the growing polypeptide chain on the polysome. Different tRNAs contain between 70 and 85 nucleotides. The most characteristic feature of tRNA is that it contains the anticodon, a sequence of three nucleotides specific for the mRNA codon sequence. There is at least one tRNA per cell bearing the anticodon for each of the 20 amino acids. The aminoacyl-tRNA (the tRNA carrying the amino acid) binds to the large subunit of a ribosome, where antiparallel basepairing occurs between the anticodon of the tRNA and the complementary codon of the associated mRNA. The specificity of this base pairing ensures that the amino acid inserts into the correct position in the growing protein polypeptide chain. During translation, the deacylated tRNA (*i.e.*, with its amino acid removed) is released from the ribosome and becomes available once again for recharging with its amino acid.

DNA-dependent RNA synthesis is the process of RNA sythesis on a template of DNA. According to the rules of base pairing, the base sequence of DNA determines the synthesis of a complementary base sequence in RNA. Assisted (catalyzed) by the enzyme RNA polymerase, the growing RNA chain releases from the template so that the process can start again, even before the previous molecule is complete. Termination codons and a termination factor known as rho-factor end the synthesis process. In certain viruses, RNA-dependent RNA synthesis occurs, with the viral RNA acting as a template for the synthesis of new RNA.

Each nucleotide in RNA contains a ribose sugar, with carbons numbered 1' through 5'. A base is attached to the 1' position, generally adenine (A), cytosine (C), guanine (G) or uracil (U). Adenine and guanine are purines, cytosine and uracil are pyrimidines. A phosphate group is attached to the 3' position of one ribose and the 5' position of the next. The phosphate groups have a negative charge each at physiological pH, making RNA a charged molecule (polyanion). The bases may form hydrogen bonds between cytosine and guanine, between adenine and uracil and between guanine and uracil.However other interactions are possible, such as a group of adenine bases binding to each other in a bulge, or the GNRA tetraloop that has a guanine–adenine base-pair.

An important structural feature of RNA that distinguishes it from DNA is the presence of a hydroxyl group at the 2' position of the ribose sugar. The presence of this functional group causes the helix to adopt the A-form geometry rather than the B-form most commonly observed in DNA. This results in a very deep and narrow major groove and a shallow and wide minor groove. A second consequence of the presence of the 2'-hydroxyl group is that in conformationally flexible regions of an RNA molecule (that is, not involved in formation of a double helix), it can chemically attack the adjacent phosphodiester bond to cleave the backbone. RNA is transcribed with only four bases (adenine, cytosine, guanine and uracil), but there are numerous modified bases and sugars in mature RNAs. Pseudouridine (Ø), in which the linkage between uracil and ribose is changed from a C–N bond to a C–C bond, and ribothymidine, are found in various places (most notably in the TØC loop of tRNA).Another notable modified base is hypoxanthine, a deaminated adenine base whose nucleoside is called inosine.

Inosine plays a key role in the wobble hypothesis of the genetic code.There are nearly 100 other naturally occurring modified nucleosides,of which pseudouridine and nucleosides with 2'-O-methylribose are the most common.The specific roles of many of these modifications in RNA are not fully understood. However, it is notable that in ribosomal RNA, many of the post-transcriptional modifications occur in highly functional regions, such as the peptidyl transferase centre and the subunit interface, implying that they are important for normal function.

RNA and DNA are both nucleic acids, but differ in three main ways. First, unlike DNA which is double-stranded, RNA is a single-stranded molecule in most of its biological roles and has a much shorter chain of nucleotides. Second, while DNA contains deoxyribose, RNA contains ribose, (there is no hydroxyl group attached to the pentose ring in the 2' position in DNA). These hydroxyl groups make RNA less stable than DNA because it is more prone to hydrolysis. Third, the complementary base to adenine is not thymine, as it is in DNA, but rather uracil, which is an unmethylated form of thymine.Like DNA, most biologically active RNAs, including mRNA, tRNA, rRNA, snRNAs and other non-coding RNAs, contain self-complementary sequences that allow parts of the RNA to fold and pair with itself to form double helices. Structural analysis of these RNAs have revealed that they are highly structured.

Chapter 5

The Genetics System

The genes are much too small to be seen individually, but in certain stages many chromosomes show lengthwise patterns of knobs or bands that indicate their ubdivision into genes. In par ticular, this happens during the early stages of meiosis. A pair of rye chromosomes at the beginning of meiosis. You notice the pattern of darkly stained knobs and faintly stained connecting strands. Thepattern in the two partner chromosomes is the same, and pairing is exceedingly accurate, each knob in one chromosome lying next to the matching knob in its partner. It appears that it is the genes rather than the chroosomes as a whole that direct the pairing process. This ability of the genes to find and attract their exact counterparts is one of their many remarkable properties for which science has not yet found a satisfactory explanation.

Later on in meiosis the pattern will be lost because the chromosomes curl up and contact into the compact bodies. But when the next generation makes germ cells, each chromosome will again show its characteristic pattern, so that a trained cytologist can recognize every one of the seve chromosomes of rye by the pattern of its knobs. Giant chromosomes withbeautiful and intricate patterns of dar and light bands are found in the salivary glands of the larvae (maggts) of certain flies. Chromosomes in the salivary glands of Drosophila larvae. They are hundreds of times aslong as ordinary chromosomes, and each of them has its own characteristic pattern of bands. They are also much thicker than ordinary chromsomes because each consists of a bundle of many strands, all with exactly the same pattern of bands. The way these bundles arise is intresting because it illustrates the extreme accuracy of chromosome replication. In the young larva, the chromosomes in the salivary glands are still thin. As the larva grows, the salivary glands grow too, but they do so in a particular way.

The cells, instead of making more cells by division, grow bigger and biggerwithout division. Their nuclei, too, grow bigger and bigger. Inside the nuclei, the chromosomes make replicas of themselves as they doin preparation for mitosis but, as there is no mitosis, the new chromosomes do not separate from the old ones. Instead, new and old chromosomes remain closely attached to each otherand each, in turn, makes more replicas. This goes on repeatedly, and so accurately does each band make another band like itself that the whole bundle appears as one single thick banded thread. In addition, the partner cromosomes in the salivary glands are closely paired as in meiosis, so that each apparent chromosome represents one pair, and thenuclei of the salivary glands of Drosophila seem to have four chromosomes instead of the eight that are visible in ordinary cells).

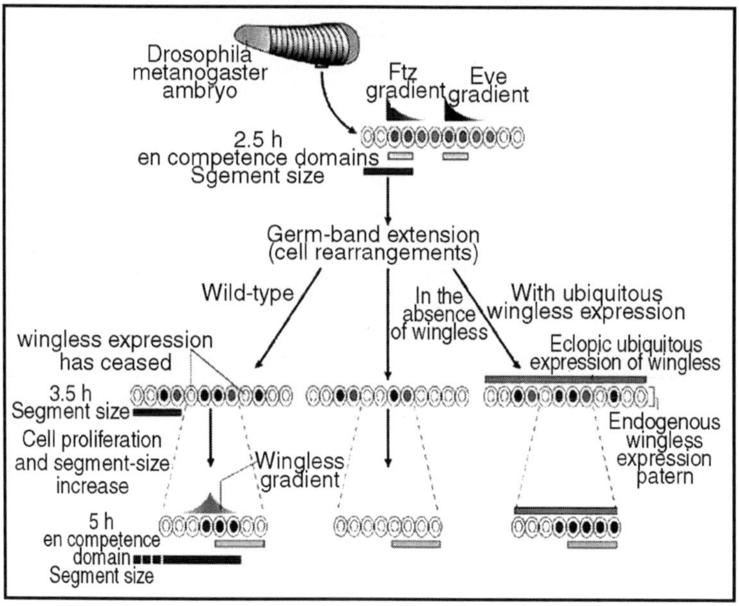

Figure 1: Wing Gene in Drosophila

The pattern of bands corresponds to the pattern of underlying genes. This can often be seen in cases where a chromosome has lost a small piece through radiation or some other injury. The accuracy of chromosome replication is such that all descendants of the damaged chromosome lack exactl the same piece, and when such a damaged chromosome gets into a gamete, for instance an ovum, individuals are formed which lack the same chromosome piece in all their cells. The left part of a normal wing of Drosophila and above it, and at a very much higher mgnification, a little section of one of the normal salivary gland chromosomes. The arrow points to a band which encloses a gene for wing shape. The right part of the same chromosome section in afly which had received a damaged chromosome from its X-rayed father or mother.

If you recall that the salivary gland chromosomes consist of two chromosome bundles formed by the two partner chromosomes, you will unerstand why only half a band is missing at the point of the arrow. Nevertheless, the lack of one of the two partner genes has been sufficient to disturb normal development of thewing and to produce the abnormal shape shown in the accompanying sketch. Flies which have lost this band from both partner chromosomes are so much damaged thatthey die as early embryos.

The left drawing represents a normally growing green maize plant and underneath, at much higher magnification, the tips of one particular chromosome pair. Thee is a fairly large knob at the end of these chromosomes. Among the genes which this knob encloses is one whose action is necessary for the production of the green substance chlorophyll that is essential for survival of the plant.

The right picture shows that, when this knob has been lost rom both chromosomes, the plant dies early as a yellow and sickly seedling. If you turn once more to the left picture you will notice a striking difference from what happened in the case of the Drosophila wing gene. Whereas this gene could not produce anormal wing without the co-operation of its partner gen, the chlorophyll gene is sufficiently effective even when acting singly; for plants with one normal and one knobless chromosome are quite normal.

Tis is not a distinction between the genes of Drosophila and maize, or between genes concerned with wing hape or chlorophyll. It is a difference in the efficiency with which genes carry out their functions; some genes are elfsufficient in this regard, others need the co-operation of their partners for effective action. We shall meet a similar difference when we deal with dominant and recessive genes. Each gene is concerned in the control of a particular developmetal process, for instance the formation of chlorophyll, the development of colour or size or, in higher animals, mental ability. Partner genes control the same process, but they may do so in different ways; thus two partner genes in peas are concerned with the control of seed shape, but one causesthe seeds to be round, while the other makes them wrinkled.

Partner genes with different effects are called alleloorphs or, for short, alleles. The term "allelomorph" signifies "diffrent shape"; according to modern ideas this may be interpreted as different chemical configurations of basically the same gene. All developmental processes are highly complex, and it is therefore not surprising that many different gene pairs are involved in the develpment of, say, an eye or a brain or a flower. Even such a seemingly simple achievement as the formation of a vitamin by a bacterium proceeds in a series of steps and is controlled by a series o genes.

In bacteria, genes controlling the same process tend to lie close together, as though this facilitated their co-operation; but in most other organisms this is not found to be so. Genes which are involved in quite different processes may lie cheek by jowl on the same chromosome, while genes controlling closely related processes may lie far apart. The position of a few of the known genes on a particular pair of chromosomes of the fruit fly Drosophila melanogaster.

Only genes concerned with the development of the eyes and the wings are represented; more genes controlling development of these organs are present on the same chromosome pair and on the three others. Some of te represented genes can be seen to act in very different ways; thus wing gene w_1 is concerned with the control of wing shape, while wing gene w_2 helps make the correct pattern of wing venation. For other genes such a distinction is not evident; thus eye genes e_1, e_2, e_3 and e_4 ar all engaged in the formation of eye colour. Yet we know from developmenal studies that each of them has a different role to play in this process.

Figure 2: The Positions of Some Wing and Eye Genes

The genes, like the chromosomes, are arranged in pairs; with the chromosomes they segregate into different gametes at meiosis and come together again at fertilization, one member of each pair being derved from the father, one from the mother. The arrangement of the genes on the chromosomes is fixed for each species, and partner genes occupy identical positions on partner chromosomes. Allelomorphs are partner genes with different effect, that is, they are genes whch occupy the same position on the same chromosome pair and control exactly the same developmental process, but they do so in different ways.

Genetic Code

Even more remarkable is that non-overlapping triples of the letters *A, C, G* and *T* which are often reerred to as codons (especially when the triples are part of a gene), may spell out (reading in the 5' to 3' direction) the order of the amino acids that form the long linear molecules known as proteins. It turns out there are 20 amino acids which are the building blocks for proteins. Since there are 64 possible triples (repeated letters within the triples are allowed) using the 4 letters *A, C, G*, and *T*, different triples can represent the same amino acids. There are also 3 triples (*TAA, TAG, TGA*) known as stop codons that do not represent (a common phrase being do not code for) an amino acid, but are involved in signaling a termination to the protein production process. One particular codon (*ATG*), which does represent one of the amino acids, is an indication for the start of the production of a protein. This system for coding proteins has come to be known as the genetic code.

Before the genetic code was fully understood, based on work by Marshall Nirenberg, Heinrich Matthaei, Har Gobind Khorana, and others, various mathematical ideas were invoked to suggest what kind of code might be involved. There is a diagram of the RNA version of the genetic code, where *U* takes the place of *T*. One approach to finding a gene is to look for a stretch of DNA which starts with ATG (the start codon) and ends with one of the stop or termination codons. This situation in a stretch of DNA is referred to as an open reading frame (ORF). However, not all ORFs correspond to a stretch of DNA which will initiate the production of a protein, and so the location of an ORF is not equivalent to finding a gene. A major problem facing biologists is how to locate those stretches of DNA that are genes in the large amounts of DNA found in a chromosome.

The way that DNA is involved in the production of proteins is not direct. The mechanism involves another helical molecule called RNA, of which there are a variety of types. (Among the types of RNA are messenger RNA, or mRNA, ribosomal RNA, or rRNA, and transfer RNA, or tRNA.) When a protein is to be produced, the DNA separates and a copy of the gene is transcribed (the process is referred to as transcription by molecular biologists) into a strand of RNA (ribonucleic acid). RNA, like DNA, uses an alphabet of 4 nucleotides *A, C, G,* and*U* (for uracil, a pyrimidine like thymine which takes the place of *T*). In a somewhat complex series of steps, a protein is produced. A very rough schematic of how genes create proteins. This schematic is often referred to as the central dogma. The original DNA, which are known as introns but which are not involved in the production of a protein, are separated or snipped out and a resulting string of exons from the original DNA is left. What is left is sometimes referred to as a coding sequence, and it contains the codons (in the letters *A, C, G* and *U*) which will create the protein, by the process usually referred to as translation.

The actual manufacture of the proteins involves structures called ribosomes. Production of proteins does not go on within th nucleus but in the cytoplasm surrounding thenucleus. One can express the codons involved in the manufacture of proteins either in DNA terms (using the letters *A, C, G* and *T*) or in RNA terms (using the letters *A, C, G* and *U*).

Implications

Many interesting quesions can be asked about the genetic code, including why the pattern of redundancies in the codeexists. While at one time it was thought that the code was totally universal, this has proved not to be true. For certain bacteria and in other situations the universality of the code breaks down. Cellular life forms are classified according to whether they are prokaryotes or eukaryotes. Prokaryotes are cellular organisms where the genetic material is not a part of a membrane enclosed nucleus, while in eukaryotes the genetic material is contained in a nucleus.

At one time life forms were classified as either bacteria or eukaryotes. A group of bacteria like organisms, found in very hot environments such as undersea vents, were initially classified as bacteria. However, when Archaea's genetic makeup was studied, its members were found to have an amazingly large number of genes that were not known to exist in other organisms. They seemed to be surprisingly different from the bacteria with which they had been formerly classified. It has become standard in recent years to list the Archaea in a separate category from the bacteria and eukaryotic organisms. Use of phylogentic trees was in part a tool that led to this change. It is tempting to believe that, because human beings are as complex as they are, other species will have less genetic material than humans. However, chromosome number (and the total size of the genome) varies with species in a way that seems to have no rhyme or reason. Thus, chimpanzees have 48 chromosomes, cats 38, dogs 78, rabbits 44, rats 42, turkeys 82, and horses 64. Perhaps when the genomes of many different and varied species are sequenced it will be possible to get insight into these wildly varying numbers.

Mendel's First Law: Segregation

After this excursion into cytology, lot us return to genetics. The chief tool of the cytologist is the microscope; it has shown him that the chromosomes, and with them the genes, segregate at meiosis and enter the gametes singly. The chief tool of the geneticist is the crossing of different strains, and we are now going to consider crosses which demonstrate the segregation of the genes at meiosis.

It will be useful for the description of these experiments if we recall some technical terms which have already been defined, namely "haploid" and "diploid" for cells or individuals which carry one or two chromosome sets, respectively, and "allelomorphs" or "alleles" for partner genes with different effects. To these terms I want to add two new ones which describe the genetical constitution of a diploid individual in regard to any particular pair of genes.

Evidently there exist two possibilities: either the partner genes are alike or they are allelomorphic. In the first case, the individual is "homozygous" or a "homozygote" in respect to the gene concerned; in the second case, it is called "heterozygous" or a "heterozygote."

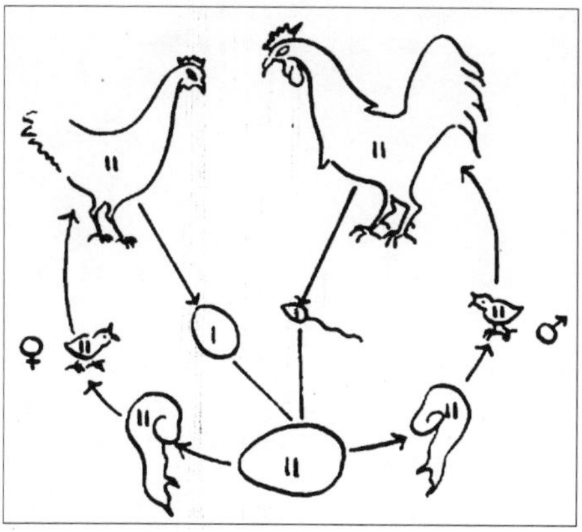

Figure 3: Life Cycle of Fowl

A pea plant with two genes for wrinkled seeds or two genes for round seeds is homozygous for either wrinkled or round; a pea plant with one gene for wrinkled seed and an allelomorphic gene for round seed is heterozygous for wrinkled and round. It will readily be seen that only heterozygotes are suitable for demonstrating the segregation of genes; for the geneticist recog nizes genes by their effects, and the segregation of like genes in a homozygote is not observable.

In higher organisms (macroorganisms) it is hardly ever possible to observe the immediate effects of segregation at meiosis; for the mature gametes fuse at fertilization and the progeny, in which we observe the effects of the genes, carries genes from two different gametes. The life cycle of the domestic fowl in terms of haploidy and diploidy. The diploid female and male form haploid ova and spermatozoa. At fertilization the gametes fuse and form the fertilized ovum, which in turn develops into the diploid embryo, chicken and adult. Thus the whole life cycle, with the exception of one cell generation, is spent in the diploid stage. The situation is quite different for many microorganisms. The life cycle of a green alga, Chlamydomonas.

Each individual of this species is a microscopically small cell which swims through the water with the aid of two whiplike hairs, or "flagellae." These cells are haploid; they have only one set of chromosomes, carrying all genes of the species in single copies. New cells are formed through division of the old ones.

In this way whole populations may be derived from one cell by repeated divisions. Such populations are called "clones." Since at mitosis both daughter cells receive exactly the same set of chromosomes and genes all cells within a clone are genetically identical.

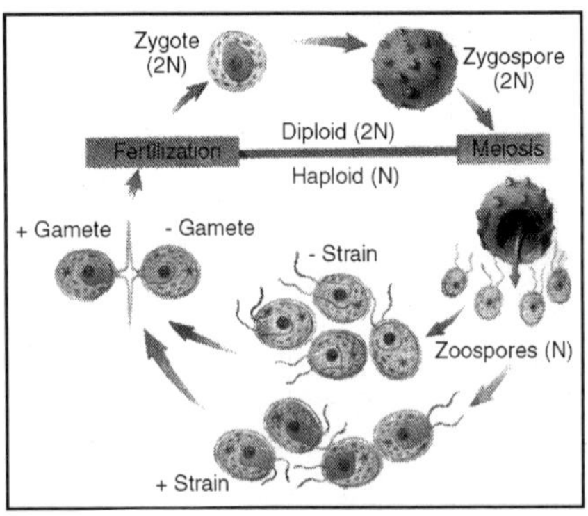

Figure 4: Life Cycle Algae Chlamydomonas

There are no visibly distinct sexes, but physiologically there is a clear distinction between sexes or "mating types," called + and–in our diagram. Cells of the same clone are always of the same mating type and never mate with each other; nor does mating occur when cells of two + clones or two–clones are brought together. But when cells from *a*+ and *a*– strain are put into the same culture vessel, pairs are fomed.

Each pair consists of one + and one–cell they fuse and form a thick skinned pore called a zygote. The zygote is comparable to a fertilized ovum in, say, a dog or a fly. It contains the chromosomes from both gametes and is, therefore, diploid. Very soon after fertilization meiosis takes place inside the zygote, and haploid gametes swarm out. These start a new life cycle by division; in fact, the gametes of Chlamydomonas do not differ from any of the other free swimming ells. Thus in this algae, as in many other microorganisms, the whole life cycle, with the exception of the zygote, takes place in the haploid stage, and segregation at meiosis is not obscured by diploidy of the progeny. Usually it is necessary to grow the gametes into clones before segregation becomes observable; but occasionally the segregating gametes themselves differ visibly from each other. Two strains (1) and (2) differ in one gene which is concerned with the development of the flagellae. While strain (I) has all the genes necessary for flagella formation, strain (2) carries a deficient allele of one of these genes and cannot, therefore, develop flagellae.

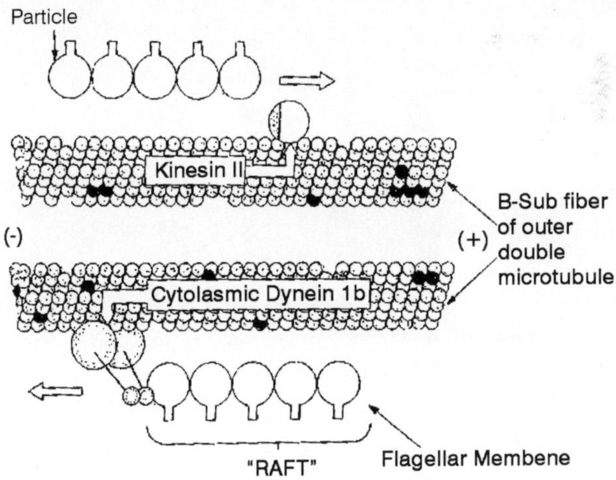

Figure 5: Absence of Flagellae in Chlamydomonas

When the two strains are crossed by bringing together + cells from the one strain with cells from the other (it makes no difference which strain provides the + cells, which the cells), a zygote is formed which carries both alleles and is, thus, heterozygous for the gene in question. At the reduction division the partner chromosomes segregate and carry the different alleles into separate cells. At the second meiotic division each cell divides once more by mitosis, so that finally four gametes are formed, two of which carry the normal and two the deficient allele. When these swarm out, it will be seen immediately that two have flagellae, two have not. This difference persists in the clones formed from the gametes.

If we let each gamete grow into a separate clone, two clones will consist of flagellated free moving cells, while two will consist of flagellaless cells which sink to the bottom of the tube.

The fact that a heterozygous zygote always produces two gametes carrying one allele and two carrying the other is so well established that it can be used, conversely, for determining whether a difference between two strains of a microorganism is due to the effects of a pair of allelomorphic genes.

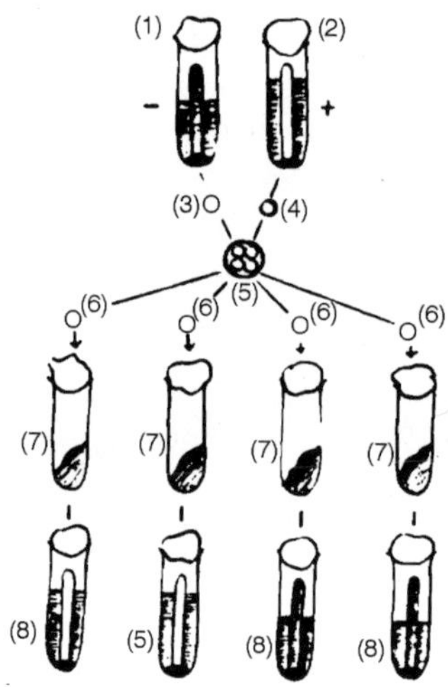

Figure 6: Segregation in Yeast for the Ability to Ferment Galactose

If this is so, the gametes from a cross between the strains will yield two clones, resembling the one strain, and two resembling the other. Yeast, like Chlamydomonas, is a unicellular organism. Many strains are haploid and can be used in experiments on segregation. Like Chlamydomonas, yeast occurs in two mating types, + and –, which are not visibly different from each other.

Yeast, as you know, ferments sugar into alcohol ad carbon dioxide gas; but strains differ in their ability to ferment various kinds of sugar. Thus some strains readily ferment galactose (one of the components of milk sugar), while others do so only very slowly and inefficiently. A simple method for detecting fermentation is to catch the developing gas in a est tube which was first completely filled with liquid and then inserted immediately above the yeast cells. If the medium contains a sugar which can be fermented, the tube gradually fills with gas; if the sugar cannot be fermented, the tube remains filled with liquid.

In the experiment shown in the medium in the top row contains galactose as the only sugar. Strain (2) is able to ferment this sugar, strain (1) is not. Cells (3) and (4) from the two strains are brought together, and a zygote (5) is formed which, through meiosis, gives rise to four gametes (6). Each gamete is grown into a clone (7) on solid medium containing glucose, a sugar which can be fermented by all strains of yeast. Finally, the four clones ae tested (8) in fermentation tubes with galactose as the only sugar. You will notice that two clones (those on the left) can ferment galactose, and two cannot.

Thus the difference between the two original parents segregates in the 2:2 ratio expected if it is due to the action of a single gene pair. On the basis of several experiments of this kind it could be concluded that in yeast the ability to ferment galactose is controlled by the action of a gene which in nonfermenting strains is absent, or is present in the form of a nonfunctioning or poorly functioning allelomorph.

The essence of what is called Mendel's first law or the law of segregation. In modern terms it may be expressed as follows: When a heterozygote forms gametes, half of them carry one of the two alleles, half the other. The basis for the segregation of the genes is the segregation of the chromosomes at meiosis. Mendel himself could not, of course, interpret his results in these terms, but he quite clearly postulated that a mechanism of this kind was responsible for his data, in particular for the famous and much overrated 3:1 ratio. This particular ratio is a rather remote consequence of the underlying gametic segregation in a ratio of 1:1. The steps by which, in crosses between higher organisms, this 1:1 ratio becomes transformed into oters, including.

In doing this we shall find ourselves using very simple statistical rules, such as are used for predicting that a coin tossed 100 times will show head about 50 times. This statistical aspect of genetics has been held against it by the Lysenko school, on the grounds that mathematics must not be applied to biological objects and that Mendel had no right to speak of a 3:1 ratio between round and wrinkled peas if some of the pods contained 6 round peas. No matter what we think about these arguments—and to a modern biologist they seem entirely futile.

Characteristics of Genetic System

X-Chromosome as Type

The chromosome combinations present in the fertilized egg, and this has brought to light some of the methods by which the chromosomes act on development and characteristics. We now Examine other effects of altering chromosomes.

We shall have to examine the effects of altering all the different classes of chromosomesýÿthe X-chromosome, the Y-chromosome, and the different autosomes. Each of these three groups (X, Y, and autosomes) gives rise, as we shall see, to a different type of inheritance. The simplest relations, from the experimental point of view, are presented by the X-chromosomes. We therefore deal first with these.

Effects on altering the X-Chromosomes

The X-chromosomes take such a course in passing from generation to generation that it is possible to follow the descendants of a particular Xchromosome (that present in the original male parent, for example), knowing in which individuals they are present, in which they are absent. In certain individuals (males, in organisms of Group I) there is but a single X-chromosome instead of a pair.

These relations make it a relatively simple matter to discover the distinctive effects of a particular X-chromosome. The effects of X-chromosomes will therefore be dealt with somewhat fully as a type of chromosomal action. Later the action of the other chromosomes will be taken up.

Abnormal Distribution of X-Chromosomes

What will happen if the X-chromosomes by accident become irregularly distributed? We know that sometimes the X-chromosomes are indeed irregularly distributed. Will the sex-linked characteristics continue to follow them; will the defective characteristics show the same irregular distribution as do the X's? Such cases in great number have been fully observed. It is found that the sex-linked characteristics do indeed follow the X-chromosomes wherever they go. This proves conclusively (if there were any possible doubt in view of the extraordinary course normally followed by such characters) that it is indeed the X-chromosomes on which the characters depend. The matter is one of importance and interest, so that it will be worth while to examine carefully certain typical cases of the result of irregular distribution of X-chromosomes.

Normal Distributions

Sometimes the body of Drosophila is yellow instead of the normal grey; this is a recessive character due to a defect in the X-chromosomes. Suppose that we mate together a female that has a yellow body and a male that has the normal grey programmes. The female has two modified X-chromosomes, which we may represent by XX. The male has its X unmodified; we may represent its cells as XY. Normally in forming the germ cells, the two X's of the female separate into different germ cells, and all the ova receive one of the modified X's. The normal male produces germ cells of two classes, X and Y.

When the germ cell X from the female unites with Y from the male, sons are produced, with the constitution XY; while X from the female with X from the father gives daughters XX. Since the yellow programmes dependent on X is recessive, the daughters XX have the normal grey programmes, like the father.

But the sons XY have only one X, and as this is the defective one, from the mother, the sons are yellow, like the mother. The normal result of such a mating therefore is that all the sons are recessive like the mother, all the daughters dominant, normal, like the father. This is what occurs when the X-chromosomes are distributed to the germ cells in the usual way.

Abnormal Distribution

In forming germ cells both X's of the female go together to a single ovum, while other ova receive no X. In a race discovered by L. V. Morgan the two X-chromosomes in the female were partly united, so that they almost invariably thus go together to one ovum, leaving other ova without an X. In this race the X-chromosomes were so modified as to produce th recessive yellow programmes described in the preceding paragraph. We therefore have an opportunity to determine the course of inheritance of the yellow and the grey programmes when the X-chromosomes are thus abnormally distributed.

Represent as before the X-chromosomes that produce the yellow programmes by X. Thus one set of ova receive XX, while another set recive no X. When the germ cells XX united with the germ cells Y from the norml father, there wee produced daughters XXY, in which both X's came from the mother, and these daughters were yellow like the mother (since no normal X was present). Sons were produced by the union of ova that contained no X with sperms that carried X; this gave sons XO, with their single X from the normal father, instead of from the mother, as is usually the case. And such sons had the normal grey body programmes of the father, instead of the yellow body programmes of the mother, as happens normally.

Thus with a change in the distribtion of the X's there is a corresponding change in the method of inheritance. If the sons receive their X from the recesive mother, as in the normal cases, they are recessive lke the mother. But when, through non-dijunction, the sons receive a dominant X from the father, they are dominant like the fther.

Similarly, in the normal case the daughters receive a dominant X from the father and are therefore dominant like the father. But in cases of non-disjunction the daughters receive both their JPs from the recessive mother, and are therefore recessive like the mother. In sum, when the X's are normally distributed, uch matings give 'criss-cross inheritance'; sons like the mother, daughters like the father. But when the X's are not so distributed, there is no criss-cross inheritance; sons are like th father, daughters like the mother. The dominant and recessive characters follow the respective X-chromosomes, whether these are distributed normally or abnormally.

Such experiments have been repeated many ties, and with other sex-linked characters. Always the characteristics follow the distribution of the X-chromosomes, however these are distributed. It may be concuded with certainty that it is the presence of X-chromosomes of a modified type that causes the appearance of the particular sex-linked characters that re manifested. The rules of distribution of sex-linked characters are the rules of distribution of the X-chromosomes.

Dominance and Recessiveness in Sex-linked Characters

As we have seen, and shall see further, many defective conditions are due to defects in certain X-chromosomes. These follow from generation to generation the distribution of these X-chromosomes. Most of these bodily defects are manifested only in individuals in which the defective X- chromosomes are the only kind present; that is, these defects are recessive.

But it is important to observe that in such cases the normal condition of the organism likewise follows the distribution of certain X-chromosomes. Haemophilia follows certain X- chromosomes. But in the same matings in which this occurs, the healthy condition of the blood follows certain other X- chromosomes. If a defective mother is mated with a normal father, the sons are defectie because they receive only the mother's defective X-chromosomes.

But in the same way the daughters are normal because they receive the father's normal X-chromosomes. In the normal individuals the X-chromosome plays a part, just as it des in the defective individuals. In the normal individuals it is healthy and supplies what is required for normal development, while in the defective individuas it fails to supply what is required. In the case of most defects this normal condition is dominant over the recessive condition; that is, when both are present, the normal chromosome is the one that prevails in its effect on the indvidual. There are defective characters, however, in which the defectivecondition is dominant or partly dominant.

In these cases, when a defective X and a normal one are present together, the defective one produces its effect, as in the case of bar-eye, described earlier. In many such cases the defect is less marked when a normal chromosome is present as well as a defective one; then the defective condition is said to be partly dominant. Thus whenever an individual bearing defective X-chromosomes is mated with a normal individual, we have both dominant and recessive characters, the course of which may be followed in the later generations the dominant character being commonly the normal or usual condition. How this works out is shown in our next paragraphs.

Tests for Sex-linked Inheritance

By observing the distribution of the dominant and recessive characters among the offspring of certain matings, a test for determining whether given characters depend on the X-chromosomes is supplied. The test consists in making what are called reciprocal crosses.

- On the one hand mate a dominant female with a recessive male;
- Also mate a recessive female with a dominant male.

Represent a dominant X-chromosome by a capital X, a recessive X-chromosome by a lower-case letter x. In the group of organisms with which we have been dealing, the female has two X-chromosomes, the male but one. The daughters receive an X from each parent, the sons an X from the mother only. A dominant female will be represented by XX, a recessive male by xo; similarly a recessive female is xx, a dominant male is XO.

The two ma tings and their results will then be represented as follows:

- XX by xo gives X + XO Dom Mother Rec. Father Dom. Daughters Dom. Sons
- xx by XO gives X + xo Rec. Mother Dom. Father Dom. Daughters Rec. Sons

Thus when the mother is dominant, the father recessive, all the children are dominant, like themother. When the father is dominant, the mother recessive, the daughters are dominant like the father, the sns recessive like the mother; this is called 'criss-cross inheritance'. Whenever reciprocal crosses give these results, it is certain that the two diverse characteristics (dominant and recessive) result from dfferences in the X-chromosomes of the two parents. 'Criss-cross inheritance' is particularly useful in showing at once that we are dealing with sex-linked inheritace.

In every case where such results are produced, further tests show that the two haracteristics follow in later generations the two different kinds of X-chromosomes wherever they go. t is mainly by the use of these tests that the many different characteristics dependent on diversities in X-chromosomes have been discovered.

Sex-linked Inheritance in Group II

Sex-linked inheritance as we havejust described it was originally discovered n animals belonging to Group I, in which the females have two X-chromosomes, the males but one (with or without a Y-cromosome). But when the tests we have just described were applied to birds and to certain other animals, it was discovered that there is a scond group (Group II) in which it is the male that has two X-chromosomes, the female but one. In th common fowl, reciprocal crosses were made between 'barred' fowls (Plymouth Rock) and black fowls (Langshan). Here the 'barred condition was found to be dominant.

The results were:

- Barred father by black mother=Barred sons + barred daughters.
- Black father by barred mother = Barred sons + black daughters.

Here the second mating gives 'criss-cross inheritance', showing that we are dealing with sex-linked characters dependent on differences between the X-chromosomes of the parents. But, in the first mating above, all the offspring are like the father instead of like the mother, which latter was the case in the matings. The father in this case has the dominant characteristic (barred), and all the offspring are dominant like the father instead of like the mother.

When the mother is dominant ('barred') the result in the fowls is criss-cross inheritance; while, in the cases before described, criss-cross inheritance occurs when it is the father that is domiant. In fact, in the birds, the males and females simply exchange roles, as compared with their roles in the organisms of Group I.

Since the roles of males and females are interchanged in the two groups, their chromosomal conditions must also be interchanged; that is, since in Group I the female has two X's, and in Group II the male plays the same role as the female of Group I, the male of Group II must have two X's, the female but one. The chromosomal conditions in the two group are shown in the above tabulation, X signifying dominant, while x signifies recessive, and o signifies the lack of X (whether Y is present or not).

Careful examination shows that this is the only way in which the results in Group II can be produced. Birds and certain moths show inheritance of the kind typical for Group II. In both these it has since been found under the microscope tha the males have indeed one more chromosome than the females.

Role of Development in X-Chromosomes

A great many characteristics, of various kinds, have been found to be the result of modifications of particular X-chromosomes, and thus to follow in later generations the distribution of the descendants of those chromosomes. Examination of a number of these is desirable, both for their own importance and for the light they throw on the functions of the X-chromosomes.

In man, the following characteristics, among others, dependent on diverse types of X-chromosomes, are known from the fact that they show typical sex-linked inheritance.

- *Haemophilia*: Lack of coagulability in the blood. This results from a serious defect in certain X-chromosomes. The existence of defective X-chromosomes having this result shows that the normal X-chromosomes play a part in supplying something necessay for producing normal blood that coagulates properly.

- *Programmes Blindness*: The fact that defectiveness in X-chromosomes causes programmes blindness shows that the normal X's play a role in producing the normally functioning eyes.

Night Blindness

- Inability to see in a poor light.
- Near-Sightedness ofcertain types.

Progressive atrophy of the muscles. The fact that defects in X-chromosomes have this effect shows that thenormal X's play a role in the normal functioning of the nerves and muscles. A considerable number of other sex-linked characters are known in man. But man is a very unfavourable organism for the study of inheritace. Yet even the little that is known of sex-linked inheritance in man shows that the X-chromosome plays a role in many diverse bodily functions.

For a more complete idea of its role, some organism that can be bred experimentally must be examined. For this urpose the fruit-fly, Drosophila melanogaster, is the best organism to select, since it has been studied more extensively than any other. In the fruit-fly we find peculiarities of the following types that are dependent on alterations in particular X-chromosomes:

- *Many different eye colours*: The normal eye programmes in this organism is a certain shade of red. Defects in different X-chromosomes result in producing, in the individuals that bear these defective chromosomes, many different shades f red, varying from deep red to a very light red, and thence to 'buff', 'ivory, and 'white'. More than a dozen different types of eye programmes are known to result from modifications of the X-chromosome. It is clear that the normal X-chromosome plays an important role in producing the normal eye programmes. Structural peculiarities of the eye, such as 'bar-eye', 'facet eye', 'furrowed eye', and the like, are known to result from modifications of the normal X-chromosome.
- *Wing modifications*: Many different conditions of the wings are known that depend upon diversities among X- chromosomes borne by different individuals. These affect all sorts of features of the wings: size, form, venation, function. The normal X-

chromosome obviously plays an important role in the full and normal development of the wings.

- *Body colours and markings*: Modifications of the X-chromosomes produce the body colours yellow, sable, tan, chrome, lemon, green, and the like, in place of the normal grey. Other changes in X alter the distribution of pigment on the body, giving the characters dot.
- *Body structure*: A defect in X produces irregularities in the abdomen, known as 'abnormal abdomen'.
- *Legs*: A defect in X-chromosomes causes abnormal development of the legs; some of them are wholly or partly reduplicated, so that the total number of legs is increased.
- *Bristles*: Modifications in X-chromosomes result in various different changes in the bristles that are scatered over the body.

Many physiological conditions and functions likewise depend upon the X-chromosomes, since they are change when these chromosomes are modified. Among these are the following:

- *Positive reaction to light*: rosophila with normal X-chromosomes fly towards a source of light. Those having defective X-chromosomes of the kind that produce a tan-coloured odydo not fly towards a source of light.
- *Weakness and short life*: Most of the, defective X-chromosmes that cause structural or other changes in the body (different eye colours, wing forms, and the like) produce likewise weakness and short life. The individuals bearing them are ess resistant to bad conditions, and live or a shorter time than the individuals that bear the nomal X-chromosomes.
- *Life and death*: Some X-chromosomes have defects that are so severe that the individuals bearing them will not live and develop, unless there is resent also a normal X-chromosome. Such defects are known as lethals. The presence of such lethal defects in the X-chromosome has the result that males bearing them die, since they have but one X; while females usually live, since they have an additional X that is often normal.

If a moher that has one lethal X and one normal one is mated to a father that has a normal X, the result is that half of the sons which receive the mother's defective X fail to develop, while the other half of the sons,receiving the mother's normal X, live and develop. All the daughters live, since they all receive a normal X-chromosome from the father. The consequence is that in such families there are twice as many daughters as there are sons. In families of 100 to 200, as occur in Drosophila, this is very striking. X-chromosomes are known to play in other organisms roles similar to those mentioned above for Drosophila, although in no other organism has the matter been so fully studied.

We may summarize what has been brought out above in the following statemets:

- Many defects that appear in individuals are the result of defects in the X-chromosomes that they bear.
- Some of these defects re dominant; they are manifested in all individuals in which the defective X i present.
- But most such defects are recessive; they are manifested only in individuals in which the defective X is the only kind present.
- When both a normal X and a defective X are present, in most cases the normal X performs the required functions, so that the individual is not defective.
- Since the female has two X-chromosomes, while the male has but one, such defects are more frequently manifested in the males. Usually one of the X's present in the female is without the defect, so that she is not defective. But when the male carries a defective X-chromosome, the defect is manifested.
- It is thus advantgeous, so far as the occurrence of defects is concerned, to have two X-chromosomes rather than one.
- This gives the female a considerable advantage over the male in these respects. There are many different types of defects due to defective X-chromosomes; almost all of them are moe common in males than in females. Some of these defects seriously injure the health, or even result in death, if the defective X is the only kind that is present. It is well known that, in general, the death rate is higher in males than in females; this is probably due to the fact that males have but one X-chromosome.
- Among the different indiiduals of a species are scattered a great number of different types of X-chromosomes having different effects on development. Many of the diverse X-chromosomes are distinctly efective, causing personal defects in the individuals that bear them. Others are not defective.

- Defects or modifications in X-chromosomes affect in different cases al parts and functions of the organism.

The Role f the X-Chromosomes in Development

What do the facts brought out in the preceding sections show as to the phyiological functions of the X-chromosomes; as to their role in deveopment? For every kind of effect resulting from defects or modificacations in the X-chromosomes, there is a corresponding (opposed or diverse)action of normal or unmodified X-chromosomes. Since defective X-chromosomes of a certain type produce in Drosophila short, ill-developed wings, it follows that the normal X-chromosomes act on the development in such a way as to give long, well-developed wings.For if we substitute the norma X for the defective one, this cause the normal well-developed wings to be produced.

Similar reasoning applies to all the defective conditions that result from defective X's; the normal X's act in such a way as to produce the corresponding normal conditions. It follows therefore that the normal X-chromosomes play a role in the production of eye programmes and structure, body programmes and structure, in the development of the legs, the wings, the bristles, in producing normal health, resistance and vigour, and in various physiological processes. We know, further, that they play a most important part in determining sex, with all that this includes of structural and physiological influence. Clearly therefore the X-chromosomes play a most important part in development. Their action is not limited to any single part of the body, nor to any single class of functions. They enter into the processes of development in such a way as to influence all parts ofthe body, all functions of the body. They begin to influence development very early, as was seen in considering the development of sex. And they play important roles in such relatively late processes as the production of eye colours.

It can hardly be doubted that they enter into the developmental activities from practically he beginning, influencing all the later processes that occur.

Autosomes and T-Chromosomes

The X-chromosomes have been dealt with because of the great advantages they offer for experimental study. It was seen that they affect development in many ways, and that in a given species there are many different types of X-chromosomes in the different individuals, causing them to show different characteristics, structural and physiological. Is the X-chromosoe typical in these relations? Shall we find similar effects in the other chromosomes? In addition to the X-chromosomes, there are the autosomes, which are present as a rule as several or many pairs. In man there are 23 pairs of autosomes: in Drosophila there are three pairs. There is, further, in many species the single Y-chromosome. The autosomes and the Y-chrmosome are distributed from parent to offspring in characteristic ways, differing from the distribution of X-chromosomes. If they affect characteristics, these differencesshould appear in a different system of inheritance. We shall examine first the autosomes.

The Autosomes: Typical Mendelian Inheritance

We have already seen that the autosomes play a role in development, in the fact that they help to determine the sex of the developing individual. In Drosophila, if the number of sets of autosomes, in relation to the number of X's, is changed, this changes the sex. Do autosomes also affect other characteristics? Do the diferent pairs of autosomes have different functions? Are there diverse types fautosomes, with different effects, in different individuals of a species, as is the case in relation to the X-chromosomes? Are there dominant and recessie effects of autosomes, as there are of X? If dominant and recessive characteristics depend on auosomes, then these characteristics must follow the distribution of the autosomes from parents to offspring, as the characteristics ependent on X follow its distribution. We must then examine the method of distribution of the autosomes whether thre are characteristics that show this method.

Method of Distribution of the Autosomes

The ditribution of the autosomes from arents to offspring is much simpler than that of the X-chromosomes.

The main facts are as follows:

- The autsomes are in pairs in all individuals. There are no individuals with unpaired autosomes, like the male of some organisms with relation to X.
- Thus the two sexes are alike as to their autosomes.
- In forming germ cells, one member of each pair of autosomes goes to each germ cell. The sperms and the ova are alike with respect to the autosomes.

  Thus the fundamental rule with relation to the autosomes is that each gem cell carries one member of each pair of autosomes.

- By union of two germ cells the pairs of autosomes are restored.
- Thus each one of the offspring carries in each pair of its autosomes one autosome from each parent.

Now consider a single pair of autosomes, such as the pair marked AA in. Suppose that in different individuals the members of this pair of autosomes produce different effects, as we have seen to be the cae with the X-chromosomes of different individuals. And suppose that, as in the case of X, one of these effects is dominant, the other recessive. How would such characteristics be inherited? This question must be examined with care. For this purpose we may take as an example the different colours found in different varieties of peas, used in the classic experiments of Mendel. The yellow seed programmes of a certain variety is dominant, and results from the constitution of the autosomes of a certain pair. The green seed programmes of another variety is recessive, owing to a different constitution of the autosomes of that same pair.

Call thetwo autosomes of tht pair in the dominant yellow race by the capital letters *AA*, while the two autosomes that give rise to the recessive green programmes in the other race may be called *aa*. The dominant parent may be designated *D*, the recssive parent *R*, When these two parents produce germ cells, each germ cell receives, of course, one chromosome for the pair. The germ cells from the dominant parent *D* have the autosome *A*, those from the recessiveparent *R* have the autosome *a*. The erm cells *A*, from parent *D*, unite each with a germ cell a from parent *R* giving offspring (zygotes) containing both *A* and *a*. These first generation offspring are commonly designated *Fi* (first filial generation). And since the characteristic carried by the autosome *A* is dominant, all these *Fi* offspring manifest the dominant characteristic *D* (yellow, in case yellow-seded peas are crossed with green-seeded peas). Next observe the result of mating together two of these individuals *Aa*, of the *Fi* generation, giving the *Fa* (second filial generation).

Each parent *Aa* gives, according to the general rule, two kinds of germ cells, one kind carrying the dominant autosome *A*, the other the recessive autosome *a*. Each kind of germ cell from one parent unites with each kind from the other. That is, half the *A* germ cells from one parent unite with *A* germ cells from the other parent, half with a germ cells from the other parent; and the same is true for the a germ cells from the first parent.

This gives in equal proportions Fa offspring (zygotes) as follows:

$$AA + Aa + a\ A + aa$$

But s the two combinations *Aa* and *aA* contain the same chromosomes, the proportions may be written *AA* + 2*Aa-f-aa* The individuals of the constitution AA and Aa contain the dominant chromosome *A*; therefore they manifest the dominant characteristic, *D*. The individuals *aa*, having only the recessive chromosome *a*, manifest the recessive characteristic, *R*. Thus in this F^2 generation, so far as manifested characters are concerned, the proportion of offspring are 3 Dominant to *i* Recessive

Thus if dominant and recessive characteristics depend on the two members of a pair of autosomes, hen a dominant individual *AA* is mated with a recessive individual *aa* the general result is bound to be that the immediate offspring (*Fi*) all manifest the dominant characteristic, while in the *Fa* generation (produced by mating two individuals from *Fi*) there will be three times as many dominant individuals as there are recessives. Now, this is the result that was discovered by Mendel in 1866; it formed the foundation for the working out of Mendelian heredity. The proportion 30 to *iR* in the *Fa* generation is commonly known as 'the Mendelian ratio'; it is one of the most characteristic proportions in this type of inheritance.

Mendel discovered that there are many different characteristics that yield this 3 to *i* ratio. If yellow and green peas are crossed, in the *Fa* generation there are 3 yellows to *i* green; if round peas and wrinkled peas are crossed, in F^2 there are 3 round to *i* wrinkled; and so for may other characteristics in the pea plants with which Mendel worked.

And since the time of Mendel these same ratios have been found for thousands of different characteristics, in great numbers of different organisms, from the pea plant and *D* roso- phila to man. The fact that inheritance gives these ratios is commonly spoken of as Mendel's law, and the type of inheritance which it exemplifies is spoken of as Mendelian inheritance. The type that gives in the *Fa* generation 3*D* to *iR* is the commonest method of inheritance, much commoner than the sex-linked method, in which the characteristics depend on differences in the X-chromosomes. This is obviously because autosomes are more numerous than X-chromosomes.

We find then that inheritance according to Mendel's law takes place exactly as it would if the characteristics concerned depended on the two members of a pair of autosomes. As will be shown later, it can be proved experimentally that such characteristics do indeed depend upon autosomes. Thus the rules of Mendelism are the rules of distribution of the two members of a pair of autosomes, one having a recessive effect, the other a dominant effect. Mendelian heredity is heredity of characteristics that depend on differences in autosomes.

Mendel did not know of the relation of the characteristics to the chromosomes. He discovered, however, certain other important facts about this type of heredity, facts which have become clearly explicable since it has been learned that this type depends on differences in autosomes. These facts must be examined. For this purpose it will be convenient first to define certain terms which are universally employed in dealing with inheritance.

By the union of two germ cells or gametes, as *A* and *a*, there is produced a zygote. If the two germ cells are alike with respect to the chromosomes or characteristics that they carry, the zygote so produced is spoken of as a homozygote; thus *AA* and *aa* are homozygotes. If the two germ cells that unite to form a zygote are unlike, as *A* and *a*, the zygote so produced is called a heterozygote, as *Aa*. The two corresponding characters, one dominant, the other recessive, that are produced by the two differing members of a pair of chromosomes are spoken of s alternative characters, or alleles, or allelomorphs. Any given individual manifests one or the other of the two alternatives, not both. For the alternative characters we shall employ henceforth the term alleles.

The characteristic manifested by a zygote is its phaenotype while its actual constitution is its genotype. Thus, two zygotes *AA* and *Aa* both manifest the dominant character; the dominant character is the phaenotype, or the individuals are said to be phaenotypically dominant. But they are diverse in genotype; one is *AA*, the other *Aa*. In the case of a homo-zygote *AA* or *aa*, the phaenotype and genotype may be held to coincide. Mendel crossed yellow peas with green peas, and found, as before stated, that in the *Fi* generation all the peas were yellow, so that yellow is the dominant allelomorph. In the F^2 there were three times as many yellows as greens. The actual number of seeds in the F^2 generation was in his experiments 8023, of which 6022 were yellow while 2001 were green, so that the ratio of dominants to recessives was, accurately expressed, 3.01 to *i*.

Mendel found that if he bred further the individuals of the F^2 generation, some of these gave different results from others. In the pea plant it is possible to allow each plant to fertilize itself, so that both kinds of germ cells come from the same parent. When this was done with the individuals of the F^2 generation, the following results were observed: The green plants (recessive) gave only recessive green offspring (they 'bred true'). Of the yellow plants (dominant), one-third gave only yellow offspring (they 'bred true'), while two-thirds of them gave both yellow and green offspring, in the proportion of 3 yellows to *i* green.

The reason for this is clear when we consider the chromosomes on which the characters depend. If *A* represents the chromosome (autosome) that gives the dominant yellow programmes, while a represents the chromosome that gives the recessive green, the original parents were *AA* and *aa*, the *Fi* generation were all *Aa*, while the *Fa* generation showed the roportions *AA-f* ^{2} *Aa* + *aa*.

The recessive individuals *aa* are homozygotes; teir two chromosomes are alike, being *aa*. When germ cells are produced by these homozygotic individuals, all these germ cells must contain the chromosome *a*, and when two of these unite in self-fertilization, of course they produce in all cases again individuals *aa*. These are, of course, recessives (green), like their parents; that is, the green parents aa 'breed true'.

Similarly the dominant yellows, *AA* (constituting one- third of all the dominants), are homozygotes. Their germ cells all contain the chromosome *A*, and the offspring produced by union of such germ cells, in self-fertilization, are all *AA*, and so doinant yellow like the parents. One-third of all the dominants thus breed true. But the individuals *Aa* are heterozygotic; that is, they have two kinds of chromosomes, *A* and *a*, in the pair.

Thus they produce two kinds of germ cells, *A* and *a*, in equal numbers. And when these germ cells unite, they necessarily give, as we saw on an earlier page, offspring of three kinds, in the proportions *AA* + 2*Aa*4-*aa* Of these, *AA* and 2*Aa* will be dominants (yellow), while *aa* will be recessive (green). Thus the homozygotes *aa*and *AA* breed true, while the heterozygotes, *Aa*, do not.

As Mendel discovered, if the heterozygotes *Aa* are mated with the recessive homozygotes *aa*, they give two kinds of offspring, *Aa* and *aa*; half (*Aa*) being dominant,the other half (*aa*) being recessive. If the heterozygotes *Aa* are mated with the dominant homozygotes *AA*, the offspring *AA* and *Aa* are all dominant, though half (*AA*) are homozyotes, the other half (*Aa*) heterozygotes. All these results are typical for the descendants of two parents different in a single pair of chromosomes, one pair being dominant, the other recessive. For reference it will be well to make a table of these results, characteristic for Mendelian inheritance.

The Results of Mendelian Inheritance

Parents: *AA* and *aa*; both homozygotic. *Fi*: *Aa*; all alike, dominant heterozygotes. *Fs*: *AA* + 2*Aa* + *aa*. Three-fourths of the individuals dominant, onefourth recessive. Half the individuals homozygotes (*AA* and *aa*), the other half heterozygotes (*aAa*). Each type of homozygote bred by itself produces offspring like itself. The heterozygotes *Aa*, when bred by themselves, produce again offspring. *AA* + 2*Aa* + *aa* *Aa* mated with aa yields again *Aa* + *aa* *Aa* mated with *AA* yields again *Aa* + *AA*. All of these results are found to hold true for thousands of different characteristics.

We may therefore summarize the situation in the following general propositions:

- When two different individuals are crossed, in many cases the differences between them follow the same method of distribution as do the two members of a pair of autosomes ýÿone being dominant, the other recessive.
- This distribution is what is known as typical Mendelian heredity.
- The two sexes do not differ in these respects. Each sex has the same proportion of dominants and recessives.
- The rules of Mendelian inheritance are essentially the rules of distribution of the members of pairs of autosomes.

Characteristics that depend on two or more pairs of autosomes.We have thus far followed the distribution of characteristics that depend on differences in a single pair of autosomes. But most organisms have several or many pairs of autosomes: the fruit-fly has three pairs, man has twenty- three

Do some characteristics depend on differences in one pair of autosomes, others on differences in other pairs? Have the different pairs of autosomes thus different functions in development and inheritance? We find that the different pairs of autosomes have indeed different functions, and that some characteristics depend on one pair, some on another. This is shown in many cases i which two parents differ in two diverse characteristics. One characteristic may then act as if dependent on one pair of autosomes, the other on another pair. As an example of this, we may take a case studied by Mendel in peas.

One of the original pea plants had seeds that were (i) yellow, and (2) round, or smooth. These, as it turns out, are both dominant characters. The other plants, crossed with those just mentioned, had recessive characters; the seeds were (i) green, and (2) wrinkled.

When these two kinds of pea plants were mated together, it turned out, as we shall see, that yellow and green behave in inheritances as if due to a difference in the two parents in the effects of a certain pair of autosomes; that is, they are alleles. Round and wrinkledbehave as if due to a difference between the two parents in the effects of another pair of autosomes. We may therefore represent the two pairs of autosomes. Representing the dominant characters by capitals, the recessives by lower-case letters, we may in the diagram represent the chromosomes giving origin to the four characters as follows:

$$A = \text{yellow} \quad a = \text{green}$$

$$B = \text{round} \quad b = \text{wrinkled}$$

The germ cells from these parents contain, of course, one member of each pair, as shown in Fig. One parent produces gametes all of the constitution *AB*, the other gametes all of the constitution *ab*. When a gamete *AB* from one parent unites with a gamete *ab* from the other, all the offspring (*Fi*) thus produced have the constitution *AaBa*, as shown at *Fi* in Fig.. Since the dominant autosomes *A* and *B* are both present, all these individuals are domiant for both pairs of characters; that is, they all have seeds that are yellow and round, as Mendel found. But all are heterozygotes-for both of these characters.

Next we mate together two of these *Fi* individuals to give the F^2 generation. In forming gametes the rule is, of course, that each gamete contains one member of each pair of chromosomes. But the two pairs, *Aa* and *Bb*, are distributed to the gametes quite independently. Thus, some get *A* and *B*, some *A* and *b*, some a and *B*, some *a* and *b*. There are thus from each parent four types of gametes, *AB, Ab, aB*, and *ab*, each type being represented by many gametes. The gametes of each of the four types from one parent unite in approximately equal proportions with gametes of each of the four types from the other parent. That is, gametes of the type *AB* unite in different cases with *AB, Ab, aB* or *ab* from the other parent; and so of each of the other gametes of the first parent. The result is to give a set of 16 combinations of gametes (some of which are alike, however, as will be seen). It is not necessary to carry further in our diagram the outlines of the chromosomes.

Inheritance Depending on Two Pairs of Autosomes

AB AB AB AB

AB(i) *Ab*(6) *aB*(7) *ab*(s)

Ab Ab Ab Ab

AB(6) *Ab*(2) *aB*(5) *ab*(8)

aB aB aB aB

AB(7) *Ab*(5) *aB*(3) *ab*(9)

ab ab ab ab

AB(5) *Ab*(8) *aB*(9) *ab*(4)

But, as will be observed, not all these 16 are diverse; thus *ABAb* is the same in constitution or genotype as *AbAB*. If we number the different combinations, and give the same number to those that are alike, as is done (in parentheses) we find that there are nine different combinations or genotypes. That is, two parents that are both *ABab* produce offspring of nine different constitutions or genotypes.

When the parents differed in but one pair of characters there were but 3 different combinations. When there are two pairs the number is $3 >$ or 9^-.

These nine different combinations will manifest the dominant characters *A* or *B*, in all cases where an *A* or *B* is present whether present once, as *A*, or twice, as *AA*); they will manifest the recessive characters *a* or *b* in all cases in which *a* or *b* is present twice, no corresponding capital letter being present. Collecting together these different types, we find the different characteristics to be manifested in the following proportions of the zygotes:

$$9AB + 3Ab + saB + 1ab$$

That is, 9 out of 16 are dominant for both pairs of characters, 3 are dominant for *A*, recessive for *b*; 3 dominant for *B*, recessive for *a*, and *i* out of 16 is recessive for both. There are thus 4 different phaenotypes.

In Mendel's experiment with peas, in which the two pairs of characters were:

1. Yellow and green;
2. Round and wrinkled; in *Fi* the proportions were 9 that were yellow and round,
3. That were yellow and wrinkled,
4. That were green and round, and *i* that was green and wrinkled.

These proportions hold generally when there are two pairs of characters connected with different pairs of autosomes. It may be observed that the proportions of the individuals of different constitution or genotype in F^2, descended from a cros between parents that differ in respect to two pairs of autosomal characters (so that the parents are *AABB* and *aabb*), can be found directly from the proportions for a single pair of characters. In *Fa*, for the single pairs the proportions are $AA + 2Aa\text{-}haA$ and $BB + 2Bb + bb$ respectively. If now we multiply together algebraically these two polynomials, we obtain the relative proportions for the different combinations produced when the two pairs are together.

These proportions are the following:

$$AABB + 2AaBB + aaBB + 2AABb + 4AaBb +$$

(1) (7) (3) (6) (5)

$$2aaBb + AAbb + 2Aabb + aabb$$

(9) (2) (8) (4),

Certain other relations among the nine different types are important, since they hold for all cases of two pairs of characters that are dependent upon different pairs of autosomes.

These are the following:

- There are four combinations (Nos. 1, 2, 3, 4), constituting one-fourth of all the zygotes, that are homozygotes for both pairs of gens. The four homozygotes are diverse in their combinations and in the characters they manifest. Any one of these when self-fertilized or mated with another like itself breeds true: that is, it produces offspring that have the same constitution as the parents.
- Four out of the 16, or one-fourth of all the zygotes, are heterozygotes for both pairs of characters. They form the class numbered (5), and constitute the diagonal from upper right to

lower left. Any of these when self-fertilized or mated with another like itself gives again the nine different types, in the proportions above set forth.

- The one-half of all the zygotes, are homozygotes for one of the two pairs, heterozygotes for the other. Four are homozygotic for *A* and heterozygotic for *B*; the other four are heterzygotic for *A* and homozygotic for *B*. In these 8 there are but 4 diverse combinations or genotypes.

If any one of these is self-fertilized, it yields 3 diverse types with respect to the pair for which it is heterozygotic. Thus, the class (6), having the constitution AABb, yields when self- fertilized three combinations in the following proportions:

$$AABB + 2AABb + AAbb$$

The above relations hold for all cases in which two pairs of characteristics are dependent on differences in two different pairs of autosomes, whatever the method of designating the characters or chromosomes. In a similar way, two parents may differ in three characters that are dependent on three different pairs of autosomes. Such parents could be represented as *AABBCC* and *aabbcc*.

Their progeny (*Fi* generation) would all have the constitution *AaBbCc*. Such *Fi* individuals of course produce a variety of germ cells, in accordance with the principles that each germ cell receives one member of each pair of autosomes, and that the different pairs of autosomes are distributed independently. Each parent *AaBbCc* thus yields the following 8 types of germ cells:

ABC, ABc, AbC, aBC, Abe, aBc, abC, abc.

When each of the 8 kinds from one parent unites with each of the 8 kinds from the other, there are produced, of course, 64 groups. If we classify the 64 individuals by the characters which they manifest, we find that there are 8 different phae- notypes manifesting respectively the following characteristics:·

ABC, ABc, AbC, aBC, Abc, aBc, abC, and *abc*

That is, some are dominant for all three characters; some for a particular two and recessive for the other; some are dominant for a particular one and recessive for the other two; and some are recessive for all three. The different phaeno- types are present in different proportions; the proportions when a large number are examined are as follows:

2*ABC* + *gABc* + *gAbC* + *gaBC* + *$Abc* + *saBc* + 3*abC* + *iabc*

The total number of different combinations (genotypes) that occur in such a case is. These combinations can be obtained in their correct proportions by multiplying together algebraically the expressions: Similar but still more complex series are obtained if parents are interbred that differ in four or more characters depending on four or more different pairs of autosomes. If the number of pairs of autosomes is called ↔, the following relations hold:

If parents differ in n autosomes:

- The number of diverse types of gametes produced by the individuals of the *Fi* generation is 2^n.
- The number of different combinations (genotypes) in the zygotes of F^2, formed by mating the 2^n gametes from *Fi*, is s^w.
- The number of different phaenotypes produced in *Fa* is 2^n.

These relations yield very great numbers of diverse combinations, in cases where the number of pairs of autosomes is large. Thus in man, with 23 pairs of autosomes, if all the autosomes differed in the two parents, the number of types of germ cells produced by the *Fi* individuals would be 2, or many millions, the number of possible diverse genotypes in F^2 would be 3, or many billions, and the number of possible different phaenotypes would be 2. (As will appear later, the number of different possible combinations is greatly increased by the exchange of parts by the members of a pair of chromosomes.)

Thus typical Mendelian inheritance, dependent on differences between different pairs of autosomes, can be treated to a certain extent mathematically. The individuals in F^2 and later generations occur in proportions represented by simple mathematical ratios.

It will be worth while to formulate the main facts that we have brought out as to autosomal inheritance:

- Some single pairs of characters, dominant and recessive, follow in inheritance the method of distribution of a single pair of autosomes.
- Some pairs of inherited characters follow the rules of distribution of two or more different pairs of autosomes, in the following particulars:
  - One member of each pair goes into each germ cell.
  - The members of the different pairs are distributed to the germ cells independently, so that all possible combinations containing one member of each pair are formed with equal frequency.

 — Each type of gamete from one parent unites with each type from he other parent, the different matings being equally frequent.

- These methods of distribution result, for the characters manifested in the Fs generation, in certain typical numerical proportions or ratios, known as Mendelian ratios.
- This method of inheritance is known as Mendelism, or Mendelian inheritance. The rules of Mendelism are the result of the rules of distribution of pairs of autosomes.
- Most characteristics of organisms are inherited in this way; many more than are inherited in the sex-linked manner. This is evidently because there is but a single pair of X-chromosomes, while there are several or many pairs of autosomes.
- Thus it is clear that, among the different individuals of a species, there are many different types of autosomes (as there are different types of X-chromosomes), producing many kinds of characteristics.

The T-Chromosome

There remains another chromosome, not included in the autosomes: namely, the Y-chromosome. Has this functions similar to those of X and of the autosomes? Does it too play a role in heredity? As we shall see, the Y-chromosome differs in many ways from the other chromosomes. Let us first summarize the main facts as to the occurrence and distribution of Y-chromosomes.

- Some organisms have no Y-chromosomes. Among those are many insects the dog, the cat, the horse.
- In some organisms it is small and appears to be degenerate. This is the case in many insects; also in man.
- In some organisms the Y-chromosome is large, but of a different form from its mate X. Such is the case in Drosophila
- In some organisms Y is not distinguishable from X inform or size, but experiment shows that it differs from X in function.
- The Y-chromosome descends from father to son only, exclusively in the male line. Normally it never occurs in females.
- Hence any characters that depend on the presence of a particular kind of Y will *(a)* be found in males only, and will *(b)* never be inherited *through* females (since Y does not pass through females).

The T-Chromosome in Drosophila

The functions of the Y- chromosome have been more thoroughly studied in Drosophila than in any other organism. Most of the facts as to this have already been brought out, in connection with our account of the functions of the X-chromosome.

We may summarize as follows the known facts as to the functions of the Y-chromosome in Drosophila:

- The Y-chromosome, although it normally occurs in males, is not required for the production of male individuals. Through non-disjunction, male individuals are produced that contain no Y.
- But the presence of Y is necessary for the fertility of males.
- Y may exist in females, as a result of accidental nondisjunction; it does not affect teir sex.
- During many years of intensive study no characteristics were found that follow the Y-chromosome.
- But recently it has been shown that the presence of Y prevents the appearance of a certain recessive character ('bobbed') that is due to a modified X-chromosome. This modified X results in the production of shortened bristles on the body of the fly. But this peculiarity does not occur if a Y is present, so that Y has the dominant normal allele for 'bobbed'. Indications of certain other ill-defined functions of the Y-chromosome have of late come to view.
- No diverse modifications or defects such as are known in great number for the X-chromosome are known in the Y-chromosome.
- Thus in Drosophila the Y-chromosome appears to have little function; it seems nearly but not quite inert.

Active T-chromosomes in certain Organisms; Relation to Male Secondary Sex Characters. In certain organisms the Y-chromosome is more active and varied in its effects. In such cases it shows in different individuals diverse modifications, comparable to the diverse types of the X-chromosome in Drosophila, and these produce diverse inherited characteristics that appear in males only. This matter is closely bound up with the question of what produces the diversity of secondary sex characters. In many animals there are special characteristics which, like the Y- chromosome, are found only in males.

These are the male secondary sex characters: horns, mane, beard, greater size, diversities in form, and the like. Since these appear only in themale, the question arises as to whether they may be due to the presence of the Y-chromosome.

With relation to this the following is the situation:

- In most cases, the presence of male secondary sex characters does ot depend on the presence of the Y-chromosomes.

This is shown by the following:

— There are animals which have no Y-chromosome, yet have male secondary sex characters: for example, the dog, the cat, the horse; also various insects.

— In Drosophila, males that contain no Y (as a result of non-disjunction) have the typical male secodary sex characters.

— In some breeds of sheep the male has horns, while the female has none. t could be suggested that this is due to he fact that the male carries the Y-chromosome, while the female does not. But in other breeds, as Dorsets, both sexes have horns, although only the male has the T-chromosome. And in still other breeds, as Suffolks, neither sex has horns, although here as elsewhere the males have the Y-chromosome.

If a female of the horned breed (Dorset) is crossed with a male of the hornless breed (Suffolk), in the Fi descendants the males have horns, although they receive their Y-chromosome from the hornless breed. They have inherited the horns through the mother.

Thus it is clear that in these cases the presence of horns is not due to the presence of the Y-chromosome. Extensive work on the inheritance of horns and horn- lessness in sheep, goats and cattle consistently gives results which show that the presence of horns is not due to the presence of the Y-chromosome, or of any special type of Y-chromosome. In all these cases (a), (b) and (c) the presence of distinctive male secondary sex characters depends in some way on the fact that the male carries but one X-chromosome, and consequently produces the male hormone. Their absence from the female results from the fact that she carries two X-chromosomes, and consequently produces the female hormone.

- But in certain fish, some of the male secondary sex characters apparently do depend on the presence of a Y of a particular type, for they occur only where that type of Y is present, and they can therefore not be inherited through the mother. That is, the Y-chromosomes of certain individuals or breeds are so modified as to produce certain definite characteristics in the individuals that contain them, while individuals that contain another type of Y do not show these characteristics.

An example of this is found in the small fish Lebistes. In a certain race, the males have a black spot at the base of the dorsal fin, while the females are without the spot. In another race of the same species neither the males nor the females have the black spot. Thus the females of the two races are alike in lacking the black spot, while the males differ. Is this difference the result of differences in the Y-chromosomes of the two kinds of males? This may be tested by crossing members of the two races. In the sptted race, represent the X and Y-chromosome by capital letters; while in the non-spotted race, represent them by the small letters x and y.

A female of the spotted race, XX, is mated with a male of the non-spotted race, xy, thus:

- Females Males
- Parents XX *xy*
- Non-spotted
- *Fi Xx Xy*
- Non-spotted

The Y-Chromosomes

This produces in Fi females Xx, and males Xy. Neither contains the Y from the spotted race, and neither is spotted. The spot is not inherited through the females of the spotted race.

Next mate the female of the non-spotted race, xx, with a male from the spotted race, XY, as follows:

- Females Males
- Parents xx XY
- Spotted
- Fi xX xY
- Spotted

This yields in Fi females Xx, and males xY. These males carry the Y from the spotted race, and they are spotted. Thus far the spot goes with the Y from the spotted race. This Y may be traced further, for example, by breeding together the non-spotted females xx with the Fi males xY, from the preceding cross, thus:

- Females Males
- Parents x xY
- Spotted
- F2 xx xY
- Spotted

This gives in the Fs females xx, and males xY. These males are again spotted; they carry the Y from the spotted race. Wherever this Y from the spotted race is found, the black spot appears; wherever the Y is from the non-spotted race it does not appear. It appears clear therefore that whether the black spot appears in the male depends upon the type of Y- chromosome that it carries. There are two types of Y-chromo- some in the species, one producing the black spot, the other not producing it. It has been shown by Winge that there are many other programmes markings in this fish that depend for their occurrence on the presence of a particular type of Y-chromosome. They occur only in the males, and only in males that have a certain type of Y. In certain other fish also the Y-chromosomes have been shown to be diverse in different individuals, giving rise to distinctive characteristics in the male. In a certain plant, Melandrium, also, there are inherited characteristics that depend upon the presence of a particular type of Y-chromosome.

There are some indications also that in man the nature of the Y-chromosome affects certain characteristics. In a family described by Schofield, the male parent in the first observed generation had webbed toes. This was inherited in the male lines only by all the males for four successive generations. It thus followed exactly the distribution of the Y-chromosome from the original male parent. It seems probable therefore that it is due to a defect or modification in the Y-chromosome of that individual. Many other cases of webbed toes or fingers are knon that occur in both sexes, so that they are clearly not due to the Y-chromosome.

On the whole, the situation as to the Y-chromosome and its relation to inheritance may be summarized as follows:

- In most organisms Y is a degenerate chromosome, lacking most of the functions manifested by the other chromosomes.

- In ome organisms degeneration has gone so far that the Y-chromosome has completely disappeared.
- In other cases it exists, but is very small.
- But in some species the Y-chromosome has effects on hereditary characters that are similar to those produced by other chromosomes. Different types of inherited characters result in such cases from diversely modified Y-chromosomes.
- Characteristics dependent on the presence of a particular type of Y-chromosome are inherited in the male line exclusively.

Three Main Methods of Distribution of Inherited Characteristics

There exist therefore three diverse systems of distributing hereditary characterstics, depending respectively on the X-chromosomes, the autosomes, and the Y-chromosomes. In each of these the rules of inheritance are given by the special method of distribution of the three types of chromosomes, taken in connection with the facts of dominance and recessiveness.

Inheritance which follows the distribution of the X-chromosomes is commonly called sex-linked inheritance. That follwing the distribution of autosomes is known as autosomal or typical Mendelian inheritance. Less important than those just named is Y-chromosome inheritance. All these methods are included under the term Mendelian inheritance, employed in a broad sense. In some organisms still other methods of inheritance exist. In higher animals and plants most inheritance occurs in one of the three methods named in the preceding paragraph.

Characteristics Affected both by Genes and by Environment

Many of the same features that can be altered by changing the genes can likewise be altered by appropriate changes in the environment. Characteristics do not fall into two classes, one exclusively hereditary (or dependent on genes), the other exclusively environmental; but any characteristic is affected both by the materials of which the organism is composed, and by the action of the conditions on these materials. Yet, as we shall see, in practice some characteristics are more readily altered by environmental conditions than are others. Certain characteristics owe most of their peculiarities to diversities among genes. Others are readily affected both by genes and by environment. Still others depend mainly on environmental conditions.

Particularly illuminating are the cases in which characteristics depend in marked degree both on genes and on environmental conditions, so that they are altered by changes in either. A number of well-known cases of this kind will be presented, selecting those that arefrom a technical point of view fully known. In that animal whose genetics is best known, the fruit-fly Drosophila, many of the characteristics that have been studied are defects or abnormalities which are sharply dependent on the presence of particular genes. Such a one is abnormal abdomen.

This appears in the fact that the abdomen is ill formed, the segments not being regular nor sharply marked off. This abnormality is found to be due to a defect in a certain gene of the X-chromosome, so that it shows sex-linked inheritance, the abnormality being dominant. Unless that particular defective gene is present the abdomen is normal. But it is dependent too on the environment.

Individuals with the abnormal gene, if developed in a moist atmosphere, have the abnormality. But if those same individuals are developed in a dry atmosphere, they have normal abdomens. To produce the abnormal abdomen, we must have first the particular kind of defective gene that yields it. But second, even if that is present, we must have, too, .the moist atmosphere for development; without it the abnormality is not produced, whatever th gene.

Another character in the fruit-fly that acts in a similar way is what is called reduplicated legs. If a certain defective gene is present in the X-chromosome, the animals show a tendency to produce double legs, or even triple or multiple legs, in place of single ones. Butthis does not happen unless the young develop at a low temperature. To produce reduplicated legs, then, there must be present a certain type of gene, and also a certain extenal condition. Unless both are present, the peculiarity is not produced. It is clear from these cases that even though an individual inherits the type of gene necessary to produce a certain defect, it may not be inevitable that he should have that defect. A special environment may correct what inheritance leaves imperfect These considerations apply to many thngs in man as well as in other organisms.

Another example of the fact that what is produced in development depends both on th genes and on the environment is seen in the production of giants in the fruit-fly. A modified gene, located in the X-chromosome, nea its left end, causes the animals (if they have no other type of X-chromo- somes than this) to become giants; they grow to nearly twice the size of the ordinary fly.

But this increase in size takes place only if the animals having the modified gene are well fed during a certain period of their early lives. If not well fed at this particular time in life, they grow no larger than the usual flies. The giant size requires to produce it a particular type of environment acting on a particular type of gene. If either condition is not fulfilled, giants are not produced.

Again, in plants, the difference between green plants, containing chlorophyll, and white ones, is in some cases due to a difference in genes. Plants with a certain kind of genes will remain white even though grown in the light. In other cases, the difference between white and green plants is not the result of a difference in genes, but of a difference in environment. Plants that live in the dark remain white, even though they contain genes that can produce chlorophyll, while similar plants grown in the light are green. That is, to produce chlorophyll, a certain type of genes is necessary, and also a certain type of environment.

Again, there are red and white varieties of primroses; the difference in programmes is inherited; it is due to gene diversities. But a certain variety produces red flowers when grown in a cool place, white flowers when grown in a warm, moist region, as in a greenhouse. The same diversity in programmes is now due to an environmental difference.

Complex interrelations of genetic and environmental action have been described by Emerson with relation to diverse colours in maize plants. In different types, the leaves, the husks, the flowers, and other parts, differ in programmes. Emerson describes and Figs six main types, under each of which are subtypes.

These are the following:

- Purple, with its subtype, weak purple.
- Sun-red, with a subtype, weak sun-red.
- Dilute purple.
- Dilute sun-red.
- Brown.
- Green, with a number of subtypes.

The interrelations of genetic and environmental factors in these colours may be summarized as follows:

- Each programmes and each subtype is constant and hereditary, when the different types are bred under like conditions.

- Each of the colours depends on many different genes, like the eye programmes in Drosophila; they may be altered by changes in any of these genes. The different colours result from different combinations of the genes.
- When the different types are crossed they give typical Mendelian inheritance: in some cases with simple ratios, in others with complex ratios. Thus far the conditions are those for typical hereditary characters; the differences are the result of gene differences, and they follow the distribution of the different types of genes in inheritance. But, also, the colours depend on the conditions to which the plants are subjected while developing. Change of certain conditions changes the colours, as does alteration of the genes. Some of the relations are the following:
- The 'sun-red' type requires exposure to the sun for production of the red programmes. Parts protected from the sun are green. In the 'green types, on the other hand, subjection to the sun does not cause the production of red programmes. Thus the difference between red and green plants is due in some cases to diversity of genes, as when 'sun-red and 'green types are both grown in the sun. In other cases the difference between red and green plants is due to environmental difference, as when 'sun-red plants grown in the sun are compared with those of the same type grown in the shade. The plants of the 'purple' type do not depend on exposure to the sun for their programmes; they are purplish in programmes whether grown in sun or shade. The degree of programmes produced by exposure to the sun depends on what genes are present. In the typical 'sun-red' variety, exposure to the sun causes a red programmes widely distributed on the plant. In the 'dilute sun-red', exposure to the sun under similar conditions causes only a little red, at the bottom and tips of the leaves, while in the 'green type no red appears. The nature of the soil affects the programmes. 'Dilute sun-red' grown in poor soil and subjected to the sun has almost all parts red; in good soil, with the same relation to the sun, it has very little red. The 'green and 'brown' types are not influenced in their programmes by the nature of the soil.
- Certain particular chemicals in the soil affect the programmes. Presence of much nitrogenous matter tends to suppress the red; if little is present the red programmes is marked.

- Storage of carbohydrates in the leaves of the programmes-producing varieties causes increase of the red programmes. The amount of carbohydrate in the leaves may be increased by experimental procedures, as by removing the growing ears, or by bending a leaf so as to break the conducting vessels; the leaves thereupon become red in the parts with increased carbohydrate. But this effect is not produced in the 'green' types; they do not produce the red programmes even in presence of stored carbohydrates.

It appears clear, therefore, that in maize different genes cause the different varieties to have different metabolic processes, different chemical reactions, some of which cause the production of colours, while others do not. Also, diverse environmental conditions likewise cause changes in these metabolic processes, inducing programmes production in some cases.Whether programmes occurs, and its type and amount, depend on interaction of genetic and environmental conditions.

Another instructive case of the interaction of genes with environmental conditions is furnished by the conditions determining the number of facets in the compound eye of Drosophila. In the normal eye there are present about 800 facets. In certain of the flies a gene at locus 57 in the X-chromosome is so modified as to cause a great reduction in the number of fully developed facets. In this condition, known as Bar-eye, there are only about 80 fully developed facets: these form a 'bar' across the eye. This Bar-eye condition is dominant; the number of facets is reduced even though one of the two genes of the pair is normal. We may call the modified gene that causes the reduction in the number of facets the Bar gene.

The number of fully formed facets, it is found, depends on whether there is present but one or two of the Bar genes, and also on whether there is present in addition a normal gene. In the males (having of course but one X), the presence of the Bar gene reduces the number of normal facets to about 90. In the female, if both the X's contain a Bar gene, the number of facets is about 65. If the female is heterozygotic, so that but one of her X's contains the Bar gene, while the other is normal, the mean number of facets is about 358. Finally, in either male or female, if no Bar gene is present, the number of facets is about 800. Later it was discovered that this Bar gene is sometimes modified in a different way, so as to cause a less marked reduction in the number of facets. This modification is known as infra-Bar. A female containing but one of these infra-Bar genes has about 716 facets; if she contains two she has about 348.

Again, by an extraordinary process, almost or quite unique in known genetics, in some cases two of the Bar genes get united by crossing-over into one X-chromosome. This condition is known as double Bar. When both of the chromosomes carry this double Bar, the number of facets is but about 25. If one carries double Bar, the other single Bar, the number of facets is about 36. Thus the number of facets in the eye depends in many ways on the number and conditions of the genes present. This may be shown in the following diagram, representing the conditions in the two X (or Y) chromosomes of any individual, with the corresponding numbers of eye facets present. In this diagram + = the normal, unmodified X-chromosome Y = the Y-chromosome of the male *B* = the Bar gene *iB* = the infra-Bar gene *BB* = the double Bar condition (two Bar genes in one chromosome) Ghr. *i*: + *iB B iB B iB B BB BB BB BB* Chr.s: + + + *iB* Y *BB* + *iB B BB* Facets 800 716 358 348 98 73 65 45 42 36 25. All these various conditions are inherited as unit-difference, sex-linked characters; they are typical hereditary conditions.

But it is found further that the number of facets in most of these conditions depends also on the temperature to which the developing animals are subjected. The numbers given above are those observed by investigators working with fruit- flies cultivated at ordinary, not precisely controlled, temperatures. If different sets of flies are kept at different temperatures, it is found that high temperatures reduce the number of facets in the individuals that carry the Bar genes, though not in the normal individuals. Krafka found the following mean numbers of facets in the various gene conditions, at different temperatures:

Females Degrees	*Bar Males B-B*	*Bar Females B-Y*	*Double Bar Males BB-BB*	*Double Bar BB-Y*	*Temperature,*
J5	214	270	5i	61	
20	122	161	33	37	
25	81	121	25′	28	
300131	40	74	‘5	‘4	

Thus the number of eye facets depends both on the genes and on the temperature. The number 61 may be produced either by a double Bar Male (*BB*–Y) at 15 degrees or by a Bar Female (*B*–*B*) at between 25 and 30 degrees. Changing either the genes or the temperature will change the number of facets.

Similar situations occur in vertebrates; one of these may be described briefly. The spotted salamander, Salamandra maculosa, is largely black in programmes, with spots or stripes of yellow. The amount of yellow differs in different cases, so that on the whole some specimens are lighter, some darker. In the lighter individuals the yellow spots tend to run together into rows. The different grades are heritable; certain stocks are regularly darker, others lighter. If darker and lighter individuals are mated, the two conditions show Mendelian inheritance, the darker condition being dominant all the Fi generation are darker, while in Fa there are three dark to one light. The difference between the two thus depends on a difference in one gene. But it can be shown also that the extent of programmes depends on the environment. If the salamander is kept continuously for a long time (a year or two) on a light background, the yellow spots increase in size: they may run together into stripes or into large yellow areas *B*, so that the animals become much lighter. If, on the other hand, they are kept for long periods on a dark background the yellow areas decrease in size; the animals become on the whole darker. The extent of the yellow programmes thus depends both on the genes and on the environment; it can be altered by appropriately changing either.

In all these cases the following general principles apply:

- What are really inherited (passed on from parent to offspring) are the genes; that is, certain materials that in certain combinations and under certain conditions give rise to certain definite characteristics.
- With the same original genes, different environmental conditions may induce the production of diverse characteristics.
- Also, with the same environmental conditins, different genes may induce the production of different characteristics.
- The same difference in characteristics that is in one case produced by diversity of genes is in other cases produced by diversity of environment.

There can be little doubt that these general statements apply to many characteristics in man. Some combinations of human genes form a individual that is a much better culture medium for the bacteria of tuberculosis than are others. A person who gets such a combination of genes will develop tuberculosis, if he lives in a region in which the germs of that disease are abundant; while another person, with a different set of genes, will not be subject to tuberculosis, even though living under the same conditions. And the person with the genes that make him susceptible to tuberculosis will not have the disease if he prevents infection by the tubercle bacillus.

In relation to such susceptibility there are doubtless a great number of grades, dependent on just what combination of genes is present. Some are extremely susceptible, others less so but still prone to the disease, and so on, up to individuals whose gene combinations are such as to give them natural immunity to tuberculosis.

What then shall be answered when it is asked whether tuberculosis is hereditary or not? One can only say that some gene combinations predispose to it more than others, so that a hereditary factor is involved. But also there is a necessary environmental factor; it is the interrelation of the two sets of factors that determines whether the individual shall be afflicted with the disease.

Similar considerations apply to any other disease, acute or chronic. Certain combinations of genes are doubtless more readily attacked by plague, by smallpox, by typhoid, by pneumonia, than others, just as certain genetic constitutions yield more readily to extremes of temperature, to exposure to the elements, to unfit food. There is no affair of life in which the gene combination borne by the individual does not play a part. But the environment too plays a part, often an overwhelmingly important part.

The relative role played by genes on the one hand, and by environment on the other, is not the same for all characteristics. Differences in certain features of the organism are mainly due to differences in the genes of which they are composed. In other characteristics, differences are due mainly to environmental diversities. There are all sorts of intermediate conditions between these extremes. We may pass in review a number of different types of characters from this point of view.

Most of the diversities in obvious physical features in most animals are the result of differences in the genes, so that they are mainly matters of inheritance rather than of environment. In Drosophila the forms of the parts of the body, of the legs and wings, the distribution of bristles, the venation of the wings, the colours of body and eyes, are all changeable through the action of diverse genes, but diversities in these features due to environmental action are relatively rare. This is the case also for such characters in the vertebrates, including man. Most of the physical characters of human beings form of their features, complexion, programmes of hair, form of limbs, and the like, as well as sex are settled mainly while the child is carried in the mother's body, when the differences in environment are very slight.

The diversities between individuals in these respects are almost entirely though not quite entirely the result of differences in the genes; they are matters mainly of inheritance. This is demonstrated by what we find in twins. The one-egg twins, having the same genes, are usually closely alike in their physical peculiarities, while the twins derived from different eggs, and so having different genes, are much more diverse with respect to such physical characteristics.

At times, of course, one-egg twins become diverse through accidents or diseases which affect one and not the other. But physical diversities in twins due to genetic differences are much more frequent than those due to environmental differences.

In some organisms the gross physical features are readily altered by environmental diversities, so that differences between individuals resulting from the different conditions under which they have lived are common. Such is the case with the habit of growth in higher plants; trees and other plants grown under different conditions are very different in form. In some of the lower animals, such as the hydroids, a similar situation is found.

It is obvious that the details of behaviour are in a high degree under the influence of environmental conditions. The particular acts performed at given moments in any given organism are determined almost entirely by the conditions under which it finds itself. Yet organisms of different genetic constitution behave very differently under effectively the same conditions, and the general pattern of the behaviour of organisms may be largely determined by the genetic constitution.

Genes and Environment in Relation to Mental Characteristics

The relative rdles of heredity and environment are of special interest in man, particularly in relation to the characteristics that influence behaviour. For the study of these matters man is the most favourable organism that exists, although for most other relations in genetics he is a singularly unfavourable object of study. In man it is possible to make detailed studies of temperament, mentality, character, a group of characteristics hardly open to examination in other organisms. There occur in man 'identical' or one-egg twins, individuals having the same set of genes, the same genetic constitution throughout. And for comparison with these there occur fraternal twins: two individuals of the same parentage, the same age and living under the same environment, but developed from two separately formed eggs and thus having to some extent different sets of genes. By comparing these two kinds of twins with one another and with unrelated individuals, opportunity is presented for comparisons of the results of similarities and diversities in genetic constitutions with those of similarities and diversities in environmental influences.

As seen earlier the twins derived from a single egg are as a rule very closely alike in most physical respects; it is this that has caused them to be called identical twins. In some cases, however, the two differ in certain physical characteristics. There appear to be two categories of causes for these differences. First, there are physical differences that result from the way in which the separation of the twins occurred. Separation of the single egg into two does not occur in the first stages of development, as by division of the one-cell stage into two. On the contrary, it occurs much later, when many cells are present and the differentiation of bodily parts has begun, and it occurs at different ages in different cases.

There is reason to believe that when an equal division occurs at a rather early stage, before the right and left halves of the body have begun to show their characteristic diversity, the two twins produced are closely alike in their physical features. At a later period the right and left halves of the embryo have begun to differ, in the way that results as a rule in the greater strength and development of the right hand and arm. At such a stage, the embryo may divide into right and left halves, with results as follows.

Of the two embryos so produced, one will have a right limb that is already somewhat advanced in development, while the left limb, formed at the plane of separation, must begin anew. The other embryo will be in the reverse situation; it will have a left limb already somewhat advanced in development, but the right limb, formed at the plane of separation, must begin anew.

In each case the more advanced limb retains its advantage and becomes the one that after birth is strongest and most used. The result necessarily is that one of the twins remains right-handed, while the other becomes left-handed. This situation is not uncommonly found in identical twins. Similar differences in other unsymmetrical parts may be produced in the same way: for example in the whorls of hair on the head. In such respects the two members of a pair often show 'mirror-imaging', presumably due in each case to the fact that the original embryo had already become a little unsymmetrical before division occurred.

From this same situation another condition often observed is believed to result. In the embryo before division, one side (usually the right) will be a little in advance of the other. After division, the twin derived from the right half continues to retain this advantage, so that it may be more vigourous than the one derived from the left half. Such differences in vigour are not rare in twins.

Apparently they may be accompanied by a psychological difference; the more vigourous twin assumes the leadership in their lives together. This reacts further on temperament and character. How far psychological and temperamental differences so produced may go is uncertain.

It appears that in some cases division of an embryo may be unequal, so that for this reason also one twin may be more vigourous than the other, with resulting psychological differences. Diversities produced in the ways above described are different in origin and type from those commonly called environmental. Under the latter term are commonly grouped diversities resulting from the different conditions under which individuals have lived. Since the extent to which differences due to the time and method of division may go, and the kind of characteristics that they may affect are uncertain, it is often not possible to distinguish these certainly from diversities due to differences of environment after division has occurred.

Belonging to one or the other class, there are sometimes marked differences in the phsyical characteristics of one-egg twins. Komai and Fukuoka have described two fifteen-year- old Japanese boys that are obviously one-egg twins; they show the usual close resemblance in form, features, colouring, and the like. But they now differ greatly in size.

One is taller by $14^{\cdot 8}$ centimetres, or about 6 inches, and his weight is greater by $10^{\cdot 4}$ kilograms, or about 23 pounds. The two were at first of the same size, the difference gradually arising during growth. The smaller of the twins is affected by diabetes insipidus; this is probably connected with his decreased growth. The larger twin is right-handed, the smaller left- handed. Whether the difference in health and growth is in some way connected with what occurred at division of the egg is uncertain. A comparable case is described by Siemens. Two sisters, clearly one-egg twins, were as usual nearly identical in most of their physical characteristics till the age of 10. At this time one of them became affected wth a severe lateral curvature of the spine, and from this time their development was very different; at 16 the healthy twin was 4-8 inches taller than the other. The origin of the defect that produced curvature of the spine and changed development is unknown.

Identical twins that have lived together sometimes show considerable psychological differences. Newman has made extensive psychological tests on fifty pairs of identical or one- egg twins, and, for comparison with these, on fifty pairs of fraternal twins. The results of these examinations have not yet been published in full, but certain data from them have been published.

By the Stanford-Binet tests, the intelligence quotients (IQJ and the 'mental age' were determined for the two sets. The two classes of twins are best compared by the differences between the scores made by the two members of each pair. Thus the average difference between fraternal twins, having somewhat diverse genetic constitutions, is nearly twice as great as that between identical twins, in which the genetic constitutions are alike.

But it is important that the identical twins that have lived together also showed differences between members of the pairs. In different pairs there was much variation in the amount of diversity. Of the 50 pairs there were five in which the two members gave identical scores in the intelligence tests. At the other end of the scale were five pairs that showed respectively the following differences (points in the Stanford-Binet scale) between the two members: 12-6, 12-9, 13-0, 13-9, 16-0.

All these five differences are considerably greater than the average difference between the genetically unlike fraternal twins. It is clear that in some cases even identical twins living together differ much in mentality. Whether the differences are the result of original diversities consequent on the method of division of the egg, or have been produced in some way by different experiences of the two, is not known.

For comparisons as to the relative effects of genetic and environmental diversities on mental characteristics, valuable data have been obtained from the examination of identical twins that have lived apart under different environments. The first case of this kind was carefully studied by Muller, and nine pairs have since been fully examined by Newman.

In each of these ten cases the twins were separated in infancy, being adopted into different families; they have then lived to adult life under diverse conditions and influences. In addition to careful observations as to character and temperament, they were subjected to psychological and temperamental tests by the best standardized methods available.

The type of results reached by these studies will best be appreciated by first looking at the detailed resemblances and differences in some particular cases. In the first such study, by Muller, there were twin sisters that had been separated when two weeks old, and that did not see each other till they reached the age of 18; from that time until the age of 30 they lived apart more than nine-tenths of the time. Physically they showed the extreme similarity in characteristics that is usual in identical twins. Both 'have always been intellectually active', 'both have been extremely energetic, capable and popular, and they have been prominent in all sorts of club work in their respective communities' (Muller)'.

Both have had two or three attacks of tuberculosis, almost simultaneously'. The usual intelligence tests gave results very closely alike for the two twins. But 'the nonintellectual testsýÿof motor reaction time, association time, "will temperament", emotions and social attitudesýÿgave results in striking contrast with those of the intelligence tests, in that the twins gave markedly different scores in all these tests'.

The differences were on the average greater than those between two individuals taken at random, and seemed 'to be correlated with salient differences in their past experiences and habits of life'. Thus this first study of such a case indicated that the different environments and experiences of the two individuals had produced a large effect on temperament, emotions, and social attitudes, but had had little effect on such matters as are tested by intelligence tests.

An illuminating contrast with these results is given by the first pair studied by Newman. The twin sisters ('O' and 'A') were born in London and were separated at the age of 18 months. One lived in Ontario, Canada, the other in London. Their environments were very different. When they were tested, this pair gave differences in those tests in which Muller's pair gave similarities, and gave similarities in those tests in which Muller's pair gave differences. Newman says: 'The twins dealt with in this paper are very different in mental capacity. But they showed great similarity in their manifestations of will and temperament, and in their emotional reactions. So this pair of twins shows that differences in the experiences undergone affect deeply the individual's mental traits: his performance in matters brought out by intelligence tests. Important points were brought out in the study of Newman's second pair, twin sisters. The two had received very different educations; one had attended school seven years longer than the other. Newman summarizes as follows the results of his study of these:

'These twins, remarkably similar after being separated at 18 months of age and unknown to each other for 19 years, have been profoundly modified by the very different educational careers.In every test of mental capacity, whether of so-called native ability or of achievement, 'G', the more highly educated twin, has distinctly the superior mind. Obviously mental training improves the ability of an individual to score well in any sort of test'. But further, 'in contrast with the great difference in mental power stands the fact that in all the tests of emotional traits and of temperament the twins gave the impression of being remarkably and unusually similar'. Here came out the effect of their identity of genes.

Newman's third pair revealed certain other facts of importance. These were two young men ('G' and 'O'), separated at two months of age. One had lived mainly in the city while the other had lived in the country. They were examined after reaching the age of 23. In native ability they seem to be nearly identical. The one outstanding difference is in their general personalities. 'G' (who had lived in the city) impresses one as more dignified, more reserved, more self-contained, more unafraid, more experienced, and less friendly. He seldom smiles, has a more serious expression about the brows, eyes and mouth. He stands more erectly with chin held in and brows drawn down somewhat over his eyes. 'O' (the other twin) is the opposite in all these respects. He is the more typical country boy, laughs readily, and is not on his dignity at all. Newman asserts emphatically that the personalities of the boys were utterly different.

Thus through the study of these first four pairs examined, of identical twins that have lived apart, it became clear that such twins may differ in many ways. It appeared that the different environments and experiences of the individuals may have a large effect on mental and temperamental characteristics. The effects were different in ctifferent cases: in some cases the twins are alike in intelligence but differ temperamentally; in other cases they are alike in temperament and emotions, but differ in mentality. In other cases the marked difference is best expressed rather vaguely as a diversity in 'personality'.

The general impression given-by study of these first four cases has been confirmed by Newman's study of six other pairs of identical twins that have lived apart. The nine cases examined by Newman were treated by uniform methods, so that they may be compared in such a way as to bring out general relations. The mean difference in intelligence quotient for these nine pairs of identical twins that have lived apart is 8'6. This is considerably greater than the mean for identical twins that have lived together, which was 5-3. So far as it goes it indicates that difference of environment has a considerable effect in increasing the mental difference between twins.

In cases I, II, IV and VIII, the difference between the twins was very marked, but is similar to the differences between the five extreme cases of identical twins that had lived together. This emphasizes the fact that it is not possible to be certain what differences are due to something that happened at the division of the egg, and what are due to later environmental differences. The mean difference between identical twins that have lived apart is nearly the same as that for fraternal twins that have lived together.

In addition to the studies of intelligence, Muller and Newman made extensive examinations designed to test other features of personality. One set was designed to test will and temperament, to distinguish the mentally quick and slow, the deliberate and careful as compared with the hasty and careless, the aggressive and forceful or the opposite, and the like. Another type of examination is designed to test emotional peculiarities: likes and dislikes, feelings of right and wrong, and so on.

The results of such tests are not readily expressible by a numerical score, such as is employed for the intelligence tests. It will be worth while, however, to attempt a tabulation in general terms of the results of the tests in the three categories for the 10 cases of identical twins that have lived apart. Thus the general picture is that identical twins that have lived apart show many different combinations of likeness and unlikeness with respect to intelligence, temperament and emotional characteristics. There are cases in which the two are closely alike in all three categories cases in which the twins are distinctly diverse in all three categories and cases in which they are alike in two categories and diverse in the third, or alike in one category and diverse in the other two.

On the whole, it is clear that other things, beside genetic constitution, play important roles in determining mentality, temperament and emotional traits. The principal difficulty in interpreting the results lies in the uncertainty as to how much is the result of the manner in which division of the egg occurred, and how much is due to later environmental differences. Both certainly play a role. For these types of characteristics it appears probable that, of the two, the later environmental differences play the greater role. It is further clear that genetic resemblances and differences play a very large role in mentality, temperament and emotional traits.

It is to be noted that in all this study, the general environment of the twins was much alike in all cases. None of them lived in different civilizations, or at different epochs in cultural development. How great a diversity in the characteristics studied could be made by such differences remains uncertain. It is further to be noted that the tests to which the twins were subjected were designed to bring out rather permanent personal traits, as distinguished from mental content and habitudes. The behaviour of human beings depends very largely upon acquired mental content, upon knowledge, and upon habitudes and these things are in a high degree dependent on the conditions under which the individuals live and the experiences to which they are subjected.

In general, it is clear that in a human population there are great numbers of different types of individuals that are diverse because their genetic constitutions are diverse. In consequence of this they differ in their capabilities, in their tastes, in their tendencies towards any particular line of action. But all these classes, so far as they do not fall in the seriously defective groups, have marked powers of adjusting themselves to many different conditions and of following many different lines of action, depending on the conditions in which they live.

In no other organism than man is so full an analysis possible of the relative roles of genetic constitution and environment in determining the mental and temperamental characteristics, for in no other organism are these characteristics highly developed.

Chapter 6

Genes and Adaptation

A species can adapt to a variable environment in one of three ways, two of which are genetic. It can retain a large reservoir of genetic variation, thus increasing the odds that under any stressful condition some members of the group are preadapted, and survive. In fact, most species in nature are genetically heterogeneous and the maintenance of the variability is assumed to be due to natural selection. Genetic uniformity is restricted to small, isolated, inbred populations. Another strategy is for a species to move in the direction of genetically programming a complex nervous system that modifies itself adaptively during development by responding to environmental stimuli.

The two modes of adaptation are not mutually exclusive and, in fact, often coexist in different proportions. We must be careful, however, not to infer genetic variability from behavioural variability, or vice versa. As an example, Ebert and Hyde selected for high and low aggression from a phenotypically intermediate wild population of mice. The range of phenotypes in the selected lines greatly exceeded that in the foundation population.

Underneath the apparent phenotypic similarity of the wild mice was the genetic potential for extreme variation upward and downward. On the other hand, those of us who work with inbred strains of mice, each composed of genetically identical individuals, are often frustrated by the amount of behavioural variability within a single strain. Unfortunately for many experimenters, the behavioural phenotype is not rigidly predetermined by the genotype.

These two strategies of adaptation appear to be consequences of different situations. A species living in a stable, predictable environment will prosper best with a fixed behavioural repertoire that is precoded in its genotype as a plan for a specialized, unmodifiable nervous system. A species living in an unpredictable environment will do better to have its genotype programme the development of a plastic nervous system with the capacity for modification by appropriate stimuli. If part of the environment is predictable, a probably universal situation, selection will favour programming fixed responses for the predictable but not for the unpredictable components of its surroundings. A complete separation of these two modes of adaptation is impossible. Thus, modifiability of behaviour by experience along with genetic variation in its efficiency has been demonstrated in insects. And there appear to be biological constraints on learning even in such large brained creatures as humans.

Typical Behaviour

This discussion of the relations between behaviour genetics and ethology begins with two questions. How uniform is species typical behaviour? How much of its variation is attributable to genetic differences? In general, members of a species share a common ethogram as well as a common structural plan.

Even though taxonomists depend primarily on structure as a basis for classification, field observers know that behaviour is often a reliable means of distinguishing between physically similar species. The most striking examples are sibling species, par ticularly common in insects, fish, and amphibians although a few examples are known in birds and mammals. These species, however, are not our primary concern here.

There is ample evidence of behavioural polytypy within good species as defined by the free interbreeding of their members. Such polytypy is likely to be associated with social organization or food gathering, and explanations are frequently given in terms of ecological pressures and adaptive response.

Three major hypotheses could explain intraspecific behavioural heterogeneity:

- Variant behaviour is transmitted from adults to young by learning;
- Behavioural variation is an expression of genetic heterogeneity within the species; and
- Behavioural variation arises because individuals develop differently depending on the environment they encounter.

Hypothesis 1 is in a broad sense cultural, as it implies the handing down of traditional modes of behaving from generation to generation.

Cultural transmission is nongenetic except to the extent that it requires a nervous system with a capacity for learning to model the behaviour of others. In humans, cultures are a major source of behaviour variation, although their prevalence within our species does not logically exclude a genic contribution to group differences.

A well known example of a group characteristic in animals is variation in food habits among troops of Japanese macaques. In one group the habit of washing food was apparently begun by one young female and was adopted quickly by her companions. It was not found in other troops living under similar circumstances. Cultural transmission of this type may be more important in nonhumans than is generally recognized, but because it has not been widely investigated from the genetic point of view I shall not consider it further.

Hypotheses 2 and 3 emphasize individualistic responses to environmental challenges rather than modelling behaviour on the actions of others. Culture is involved only as a part of the environment in which development takes place. The difference between the hypotheses lies in the role of genetic variation. Hypothesis 2 places a premium on genetic heterogeneity as a means of ensuring that some individuals will be able to survive and propagate in environments that deviate markedly from the species optimum. This is clearly the classical concept of natural selection, and it is supported in a general way by the universality of genetic variability in natural populations.

Hypothesis 3 does not deny the role of natural selection in the past, but for the present it emphasizes the lability of phenotypic development and adaptive responses to particular environmental conditions. For adherents of this view, behavioural variation it within a species is more a matter of differences in individual life histories than in individual genotypes. Actually, the three hypotheses are not mutually exclusive. All are supported by evidence, but their relative importance varies according to circumstances.

Behavioural Polytypy

An extensive literature deals with variation in the behaviour of domesticated and laboratory animals, house mice, rats, dogs, fowl, and fruit flies. In these species it is often possible to separate experimentally the effects of genetic and environmental factors. Genetic variation has been demonstrated for activity level, emotional reactivity, eating and drinking, learning ability, and social behaviour. In some domestic species, selection for behaviour has produced breeds with special aptitude for certain tasks. These breed differences persist even when animals are reared as nearly as possible in the same manner.

As an example, Fuller trained five breeds of dogs to walk on a lead from their outdoor living quarters to a laboratory and found striking differences in their reactions. Shetland sheep dogs crowded against their guides, a pattern that strongly resembled the herding of sheep by their working relatives. Basenjis, with a short history of domestication, struggled more violently and bit their leash more frequently than any other breed.

Beagles, who have been selected to track small game and to give voice as they follow a trail, were noteworthy for their mournful vocalization. None of these behaviours was restricted to a single breed, and none was completely absent from any breed. Their frequencies and intensities differed enough, however, to define significantly different breed profiles. Crosses between Basenjis and cocker spaniels, breeds with widely differing profiles, yielded hybrids that were intermediate to their parents on all characteristics except vocalization, which was higher than that of either parent. The occurence of stable, intraspecific behavioural variation among populations is ethological polytypy. It is not restricted to domesticated animals. Polytypy has been reported within single taxa whose range extends over a wide territory that covers different environmental conditions. Examples include baboons (Papio anubis), vermets (Cercopithecus aethiops), and langurs (Presbytis entellus), whose social organization varies greatly over their ranges and appears to be correlated with the nature of the habitat.

Hendrichs has recorded variations in the social organization of three species of antelope, diddik (Madoga kirki), reedbuck (Redunca redunca), and waterbuck (Kobus ellipsiprymnus), that are associated with population density, that is, in turn, presumably regulated by the supply of food and other resources. Stacey and Bock report that acorn woodpeckers (Melanerpes formicovorus) that ordinarily live in permanent groups and defend year round territories deviated from this pattern in one location. Here, they mated in pairs and migrated in early winter. Old field mice (Peromyscus polionotus) from different geographical areas were found to have identical patterns of sexual behaviour, but these varied in quantitative aspects such as latency, frequency, and duration of the components that make up the total mating act.

There is no evidence that these polytypies are attributable to genetic heterogeneity. In fact, many of the cited authors generally ascribe these variations in social organization and behaviour to environmental pressures. The assumption, not always explicitly stated, is that given the same circumstances any member of the species could adopt either of the alternate patterns. However, there have been few tests to disprove a genetic basis, and it would be difficult to make one with many of these species. It is possible in species that can be maintained in large numbers in captivity and bred systematically.

Important contributions to the genetic analysis of ethological polytypy have been made by King and his students. The genus *Peromyscus* includes many taxa that have been classified at the subgeneric, species group, species, and subspecies level in ascending order of closeness of relationship. King, Price, and Webber compared animals from nine taxa on five motor behaviours: swimming, running, climbing, digging, and gnawing. They were searching for correlations between behavioural similarities and closeness of relationship as judged by taxonomists.

Secondarily, they also looked for correlations between behaviour and habitat. For example, burrowing species might be expected to excel in digging, woodland species in climbing. In all tests the only motivation to perform any of the actions was to escape from an unfamiliar place. Differences among taxa were found on every task, but the expectation that the degree of similarity would correlate with the degree of relationship was not fulfilled. Surprisingly, the behaviour—habitat correlations were no better.

Although the results of this study were disappointing, King rethought the problem and carried out other experiments. Instead of comparing taxa on specific motor acts in an artificial setting he designed a set of tests to measure more general behavioural characteristics: speed of adaptation, reflex responses, abstractive and strategic behaviour. Again the subjects were *Peromyscus* taxa of varied degrees of relationship and similarity of habitat. The results of this study lead to a revised view of behavioural polytypy as related to genetics. All taxa of *Peromyscus,* and many other “mice,” are roughly similar in form and are capable of “rodentlike behaviour” such as gnawing, grooming, and nest building. The major differences among them lie in their modes of organization of activities in new situations as they search for food, mates, or shelter. Some taxa become specialized through fixation of genes favouring survival under typical natural conditions; others retain their juvenile plasticity into adulthood and adapt as best they can to situations as they arise. King finds evidence that both Hypothesis 2 (fixed genetic programming) and Hypothesis 3 (flexible neural programming) apply to *Peromyscus.*

In a somewhat similar experiment with wild *Mus musculus,* captured in three ecologically distinct areas and reared similarly in a laboratory, Plomin and Man osevitz found differences in open field activity, defecation, climbing, running wheel activity, and nest building. They attributed the differences to genetic factors, although controls for maternal effects and cross breeding experiments were not included in their study. A repetition with more subjects and additional controls would be useful.

In muroid rodents, male copulatory behaviour has been studied from a genetic point of view. Dewsbury (1979b) reported basically similar factor patterns in laboratory rats, the descendants of a wild population of house mice, and deer mice. In spite of the factorial similarity each of these species has a distinctive style of courtship. In more closely related taxa, such as subspecies, male copulatory behaviour is much more uniform and for practical purposes qualitatively identical. However, significant quantitative differences in the latencies and durations of phases of mating were found in 11 of 14 measures. Subjects were three subspecies of *Peromyscus maniculatus: P.m. bairdii, P.m. blandus,* and *P.m. gambeli.* Cross fostering of deer mice on house mice produced minor quantitative effects on latencies, but, according to Dewsbury: "basic motor patterns in cross fostered mice were identical to those of normally reared animals."

A particularly interesting instance of behavioural polytypy occurs in the squirrel monkey (*Saimiri sciurus*). Two populations of this species differ markedly in the phonic characteristics of their extensive vocabulary of calls. Externally the populations are distinguishable by facial markings. The Roman arch type found in Peru, Costa Rica, and Panama, and the Gothic arch type resident in Columbia and Guyana also differ in their karyotypes.

That the dialects differ for genetic rather than cultural reasons was demonstrated by rearing infants with devocalized mothers and finding that they conversed in the dialect appropriate to their chromosomes and emitted these calls in their appropriate context. A personal communication from Newman adds that new research has shown that isolation peeps of these two forms differ so much that analysis of sonograms permits classification of individuals with 99per cent accuracy. The recorded IPs of eight F_1 hybrids were matched with their phenotypic appearance. Five hybrids with Roman arches emitted typical Roman IPs.

The IPs of the three Gothic type hybrids were variable, one strongly Gothic, one strongly Roman, and one mixed. These observations are insufficient to hypothesize a mode of inheritance, but apparently vocalization patterns are associated to a degree with physical characteristics.

The traditional genetic procedures of crossbreeding and selection. Most knowledge of the formal genetics of social behaviour comes from experiments with selected breeds or inbred lines of house mice, Norway rats, dogs, *Drosophila*. Can results obtained with these long domesticated animals lead to principles that can be extended to their wild relatives?

Lorenz asserted that results from animal strains that have been bred for a long time in captivity are not applicable to free living animals: "because captivity changes all hitherto effective selective factors in so profound a way that serious changes must be expected in the genome of the stock after only a few generations." Recent research indicates that Lorenz was too sweeping in this statement. Domestic albino rats transferred from their protective laboratory environment to outdoor pens survived severe northern winters, maintained a stable population, constructed and lived in burrows indistinguishable from those of wild rats, and in general behaved like their wild cousins.

Given their freedom, the albinos were able to cope with stresses that their recent ancestors had never encountered. Of course, in a completely feral state their impaired vision and conspicuous colouration would make them more vulnerable to predators. The important point of Boice's study is that there was no evidence of a loss of species typical adaptive behaviour. Disuse of skills does not by itself lead to their deletion from the behavioural repertoire. In a similar vein, Miller made extensive observations on Peking ducks that have been separated from their wild mallard ancestors for centuries. Pekings differ from mallards in important characteristics: They are too heavy to fly, and they lack the mallard's striking sexual dimorphism in plumage. Yet, when observed in a natural setting during three breeding seasons, they performed all the species typical displays that have been reported for mallards.

Miller concluded that his data: "offer no justification for the view that domestication has a 'degenerative' effect on social courtship behaviour patterns, but are rather supportive of the view that the customary nonoccurence of these displays is a function of the inhibitory or unfavourable environmental context in which the domestic birds usually find themselves." The observations of Boice and Miller do not negate the existence of behavioural differences between domesticated animals and their wild progenitors but they support the idea that many such differences are the results of selection for a characteristic desired by their breeders: docility in laboratory rats, and flightlessness in birds that are intended for the dinner of their custodians.

Some changes accompanying domestication probably are the effects of the loss of dominant alleles through inbreeding. Lynch found that inbred wild house mice, compared with random bred animals from the same foundation stock, had smaller litters, bore lighter offspring, and constructed poorer nests. Random bred wild mice reared in a laboratory were not larger, more fertile, or better nest builders than standard inbred strains. In fact, their nests were smaller than those of the heterogeneous HS stock produced by first intercrossing eight inbred strains and then continuing with a random breeding system. Lynch explains the hypertrophy of a behavioural trait related to fitness in a stock synthe sized from inbred strains that built only normal sized nests by postulating that some deleterious recessive alleles present in wild mice were eliminated during inbreeding. She believes that the inbred lines that survive the inbreeding process are good material for analyzing the genetic architecture of a species.

Her view is contrary to that of Smith who found that the heritability of shuttle avoidance learning was much lower in lines of mice selected from a wild stock than in inbred strains and heterogeneous stocks. The relevance of shuttle box avoidance to biological fitness. There is considerable evidence that good performance in the apparatus depends less on learning ability than on the prepotent response to alarm signal, freezing or running. Domestic rats have been selected for docility, and this may indeed disqualify them for inferring selection pressures in nature that are related to escape from shock. Some of the changes in behaviour that accompany domestication seem to be the effects of inbreeding rather than of conscious or unconscious selection. Connor compared wild house mice reared in a simulated natural environment with laboratory reared wild mice and with three standard inbred strains. Wild mice, irrespective of their habitats, were more aggressive towards intruders, vocalized more when handled, and were more difficult to recapture when they were freed in a room.

In a second experiment, Connor followed behavioural changes over 10 generations in random bred wild mice maintained in a natural or a laboratory habitat. A third group from the same original population were inbred by brother sister matings. No behavioural differences were found between the laboratory and the seminaturally reared random bred stocks. Maternal effects were looked for in a cross fostering study, but none were found. In contrast, the inbred wild lines were less aggressive and easier to recapture. The evidence is convincing that we should not ascribe intraspecific variation in behaviour exclusively to genetic or to environmental factors without evidence from experiments designed to separate their effects. Price, on the basis of his studies of the modification of activity by mode of rearing in wild and semi domesticated *Peromyscus maniculatus bairdii* came to the following conclusions.

The genetic requirements for a behavioural character should not change during breeding in captivity unless:

- It is selected for deliberately; or
- It is selected for unwittingly because it improves fitness under the conditions of captivity. These conclusions are encouraging. By a judicious mixture of field observations with controlled breeding and rearing in captivity, we someday may be able to determine the extent to which natural variability in species typical behaviour is a function of genetic differences. An example is the work of Moss and Watson, Sociobiology. Genetic heterogeneity is an absolute necessity for evolution; we know that it exists, but we do not know how important it is with respect to differences in behavioural fitness.

Genotype Environment Interaction

Neither genotype nor environment is sufficient to determine a behavioural phenotype. Without a genotype there is no organism. For a zygote or parthenogenetic ovum to develop into an adult there must be an appropriate physical and biological environment. Ethologists are especially interested in the biosocial factors that shape the development of behaviour. Indeed, the developmental process is often described as a genotype—environment (GE) interaction. The general phenomenon of joint involvement as GE coaction, and to confine GE interaction to situations in which a specific environmental change affects organisms differently depending on their genotypes. GE interactions are detected by exposing two or more genotypes to two or more conditions (*e.g.*, stimulated versus nonstimulated) and comparing the behavioural effects.

Broadhurst and Jinks distinguished two ways in which genes might affect development:

- They could influence its stability, particularly during early stages, and effect processes that could potentially influence an individual's behaviour for life. A tendency for phenotypic development to proceed on a stable, species-specific course regardless of ambient conditions is called developmental homeostasis.
- At any stage of life, genes and their productions may influence the capacity of an organism to adjust to changes in its environment such as removal from a familiar habitat. Broadhurst and Jinks' distinction between stability and change is somewhat arbitrary, but it is still useful. In general, stability refers to processes during early development that have generalized, permanent effects; change involves behavioural adaptations to a specific situation, with less carry over to other conditions.

GE interaction takes several forms. The most extreme is disordinal interaction where the effects of an environmental change are opposite in direction depending on the genotype of the subjects. Demonstrates that in a 2-genotype-2-treatment experiment, one could come to completely different conclusions regarding interaction depending on one's choice of treatments. An organism must receive an appropriate exposure to some stimuli in order to develop optimally. The low end of the environmental scale corresponds to extreme stimulus deprivation; the upper end to severe stress. Strains A and B differ in their optimal stimulus requirements.

A requires less than B, but it is also more easily damaged by high levels. Suppose now that we conduct four experiments with these strains choosing only two values of stimulus intensity for each. The chosen values are indicated by x's on the abscissa. Experiment 1 yields a clear interaction; strain A is helped by stimulation, strain

B scarcely at all. Experiment 2 produces a marked disordinal interaction. Strain A is impaired as stimulation increases; strain B is improved. In Experiment 3, both strains profit from additional stimulation, and there is no interaction. Finally, in Experiment 4 both strains are impaired as stimulation increases.

Thus the nature of any interaction found may be a matter of the choice of experimental procedures rather than evidence for fundamentally different ways of responding to a stimulus change. The remedy, obviously, is to employ a wide range of treatments in any experiment where GE interactions are likely to occur.

Unfortunately, there are few experiments extensive enough to provide a test of the generality of the threshold hypothesis. They were placed separately in cages where they had no physical or visual contact with other dogs or humans. Generally the period of isolation was 12 weeks; in some experiments isolation was interrupted periodically by brief placement in an arena where subjects could interact successively with a human handler, with another puppy of the same age, and with a rubber ball or swinging toy.

From the results of a series of experiments it was concluded that the deleterious effects of early isolation are primarily attributable to an overwhelming fear reaction when an adolescent puppy suddenly faces a complex environment that is totally unfamiliar. Isolated puppies that were sedated with drugs at the time of first emergence from isolation performed almost like normally reared individuals, demonstrating that their motor and perceptual development had been little impaired by living in the small chambers.

In most mammals reactive capacities develop gradually from infancy to maturity and the environment that must be dealt with at different ages is matched with the infant's or adolescent's perceptual and motor capacities. Rather small interruptions in isolation were sufficient to counteract the effects of experiential deprivation. Puppies who had 15 minutes per week in our test arena developed social and manipulative skills almost as well as pet reared animals with several hundredfold more experience.

A number of our experiments were directed towards defining breed differences in the intensity and duration of the effects of experiential deprivation with beagles and wirehaired fox terriers as the subjects. Beagles have a history of selection for olfactory tracking of small game animals. Litters housed together in group pens seldom fought and dominance hierarchies within litters were weak. In a familiar laboratory setting they gave the impression of relaxation, but they were somewhat shy in unfamiliar surroundings. Housed in the same way, the terriers engaged in fierce battles and formed strong dominance hierarchies within litters. When handled in tests their muscle tension was high, and they were more likely to attack than to retreat from a threat. Thus, it was not surprising that the effects of isolation were more striking in beagles than in terriers. In one experiment, beagles, transferred to the test arena in a small cage, often failed to explore the new environment; terriers rapidly overcame an initial hesitation and reacted appropriately to objects, people, and other puppies. In another set of experiments, beagles and terriers who had been isolated from weaning to 15 weeks of age were compared with pet reared siblings. Over a period of 5 weeks each subject was observed 25 times in the arena and records made of behaviours such as responding to a handler's call, following a moving person, playing with toys, and interacting with another puppy of the same age.

A quantitative scoring system was devised for two aspects of behaviour, response intensity to key stimuli and activity index. The response indices of pet reared (control) beagles and terriers were practically identical. In both breeds, isolates were inhibited in responding during early tests. Intensity scores rose with repeated exposure to the arena; the rate of increase was much higher in terriers. At the end of the study isolated beagles were still inhibited compared with controls.

On the second measure pet reared terriers were more active than similarly treated beagles throughout the observations. During early tests, isolates of both breeds were inactive, as had been anticipated from previous experiments and, as expected, activity increased in both breeds with repeated trials. Beagles were approximately equal with their controls by the fifth week. Terriers, however, were significantly more active than their controls by the second week.

At first glance this might appear to be a disordinal interaction. Experiential deprivation makes terriers more active; it makes beagles less so. In terms of overt behaviour this is certainly true. But there are other ways of looking at the data. In both breeds isolation produced an initial aversion to the arena, reduced activity, and decreased the intensity of response to the test stimuli. Both breeds recovered with experience, but the terriers improved more rapidly and more completely. To be sure, the isolated terriers overshot and became more active than their controls; isolated beagles were less active than their controls. In a sense, however, isolation acted similarly on both breeds in exaggerating the characteristics that differentiated them when they were raised conventionally. Normally active and investigative terriers became more so. Deliberate and cautious beagles became hesitant and timid. It appeared that isolation had made both breeds more emotionally reactive, but they expressed the change in divergent genotype typical patterns.

Genetic Architecture and Behavioural Phylogeny

Ethology is grounded in evolutionary concepts. Roe and Simpson's *Behaviour and Evolution* brought together the views of many distinguished scientists and is still an excellent introduction to behavioural phylogeny. The value of the comparative method in deducing the phylogeny of innate behaviour patterns, particularly communication by sound, posture, and gesture, was emphasized earlier by Lorenz.

However, there are problems with the application of evolutionary theory to behaviour. The validity of the very idea of behavioural phylogeny has been criticized on the ground that it is impossible to distinguish between similarity due to common ancestry (homology) and that attributable to common selective pressures (analogy). His warning is appropriate, but it need not lead to the abandonment of the consideration of the role of behaviour in evolution. In this presentation we are not concerned as much with homology and phylogeny as with the role of behaviour as a component of fitness and, therefore, as a directive force in evolution. Consideration of this topic requires a brief outline of theories relating the genetic architecture underlying variation in a behavioural character to the character's role in natural selection.

Mendelian Approaches to Behaviour Adaptation

We can distinguish between approaches based on one or two loci (Mendelian analysis) and those based on biometrical analysis of quantitative variability. The latter type of variation is considered to depend basically on Mendelian principles, but the numerous loci that are involved are not individually identifiable. Mather called these polygenes and we now apply the term polygenic inheritance to many behavioural traits. Both the Mendelian and the biometric approaches are important to behaviour genetics. Benzer's studies of mutants with well defined behavioural peculiarities have advanced our knowledge of neural development and function in *Drosophila*. The neurological mutants of mice have contributed to our understanding of brain development and function in mammals. Coat colour mutants, such as albinos, have been studied frequently by behaviour geneticists.

Albinos do differ from pigmented animals in many aspects of behaviour, and even normally pigmented carriers of the albino allele are impaired in some phases of development. However, neurological mutants and albinos are so rare in natural populations that they cannot be classified as true polymorphisms. We know that genetic polymorphisms are extremely common and the possibility that some of them have behavioural significance is well worth investigating.

An individual, the target of selection, is not a mosaic of characters each of which is the product of a given gene. Rather, genes are merely the units of the genetic programme that govern the complicated process of development, ultimately resulting in the phenotypic character. To consider genes as independent units is meaningless from the evolutionary point of view because the individual as a whole is the unit of selection. To regard genes as independent units is meaningless from the physiological viewpoint because genes interact with each other in producing the various components of the genotype.

For Mayr, the genotype has properties not deducible from the sum of its components. I agree. But he also acknowledges the possibility of switch genes where an allelic substitution at a single locus may shift the phenotype so that the direction and intensity of selection pressure is modified. A familiar example is industrial melanism, the replacement of light coloured variants of several species of moths by darker variants in areas where pollution has altered the colour of the bark of the trees on which the moths rest.

How much evidence is there for switch genes that change behaviour drastically enough to influence evolutionary processes? Sociobiologists seem to assume that such genes are abundant enough to be important. Trivers writes: "Assume that the altruistic behaviour of an altruist is controlled by an allele (dominant or recessive) a_2, at a given locus, and that there is only one alternative allele, a_1, at that locus and that it does not lead to altruistic behaviour." Simple allelic substitution could switch such complex behaviours as altruism and selfishness or change the mating system of a species from polygamy to monogamy. It is hard to imagine how such a switch gene would operate physiologically.

Of course, any allelic substitution that affects sensory, neural, or motor functions will have some effect on behaviour, but a single mutation seems unlikely, by itself, to shift behaviour into promising new channels. However, there are reports suggesting that some genetic polymorphisms have behavioural correlates that affect selective processes indirectly.

Krebs, Gaines, Keller, Myers, and Tamarin found large fluctuations in the frequency of alleles at the transferrin (*Tf*) and leucine aminopeptidase (*LAP*) loci in the vole, *Microtus pennsylvanicus*, that were associated with changes in population density. They believe that the gene frequency changes over time were indicative of an association between the alleles at these loci and the ability to survive and propagate in stressful situations. Specifically, they hypothesized that females heterozygous at the *Tf* locus were apt to migrate during periods of increasing population density. Homozygotes, particularly for the Tf^{c} allele tended to remain in their original territory. The implication is that the *Tf* locus has somehing to do with social interactions, but, to my knowledge, this relationship has not been verified by direct observation of individual encounters. Krebs *et al.*'s data were obtained from field experiments supplemented by genotype determination of captured animals. A group of immuno-geneticists have applied a very different approach to the genetics of social behaviour. Andrews and Boyse claim to have mapped two loci related to mating preference in house mice. They hypothesize that one of these, *Ris* (recognition of identity signal) provides a cue, probably olfactory, by which males can distinguish preferred females. *Ris* is assigned a location in the *Qa-Tla* region of the major histocompatibility region of the mouse. Another locus, Rir (recognition of identity response) is assumed to determine whether male response is positive or negative.

The genetic procedures are complex and ingenious, but I have reservations regarding the validity of their conclusions. Males were given five to six repeated tests to determine their sexual preference. The significance of deviations from the null hypothesis (random choice) was apparently computed on the basis of the total number of tests rather than on the number of individuals. The receptivity of the rejected females was confirmed by pairing them with other males, but one cannot exclude thepossibility that factors other than identity recognition by males bias the possibility of a female being bred. In spite of these experimental problems, the hypothesis that identity recognition based on cellular characteristics is important in mating choice has evolutionary implications. More strinent statistical analysis and more adequate behavioural controls are needed.

Isolation between closely related taxa need not imply a behavioural barrier. An example that involves a physiological difference with a relatively simple genetic basis has been reported in the lacewing flies, Chrysopa cornea and C. downsei. In the laboratory these species interbreed readily. In nature, crossing is not observed. C. carnea produces three generations per summer; C. downsei produces one, then enters diapause. Standard Mendelian crosses were made between the two species and the occurence of diapause in the hydrids was observed. The data are consistent with control of diapause by two loci. The C. downsei phenotype (diapause) requires recessive allele homozygosity at both loci. This case should probably not be called behavioural isolation in the strict sense, as the two species will mate when they are artificially manipulated. It does demonstrate that relatively simple genetic systems may have substantial evolutionary consequences.

Quantitative Inheritance

Let us now consider deductions that can be made regarding behavioural evolution from knowledge of the genetic architecture that regulated quantitative characteristics. It is well known that variation in such traits is largely polygenic. The theory relating genetic architecture to the evolution of behaviour has been explicated by several investigators. Essentially, stabilizing selection for an intermediate expression of a trait, and directional selection for high or low values involve different genetic strategies. Stabilizing selection requires either additivity of gene effects, so that a heterozygote is midway between the two homozygotes; or a balance between dominant genes at different loci, some with increasing, some with decreasing action on the phenotype. Traits with such a history respond well to selection in the laboratory and are said to be highly heritable.In contrast, selection for the extreme manifestations of a trait favours alleles with dominant effects in one direction only. Such a system ensures that the favoured alleles will be effective in a single as well as in a double dose. Over time, recessive alleles will tend to disappear from the selected population, and the heritability of the trait will become low. Clearly, it is most efficient for a species to have those traits most directly related to fitness regulated by dominant genes.

Falconer noted that in domestic animals heritability is lower for traits such as fertility and viability, than for traits desired by breeders for purely economic considerations.

Accepting these principles, one can work backward from determination of the genetic architecture of a trait to a reconstruction of the role of that trait in natural selection. High unidirectional dominance is evidence that extreme values of that trait increase fitness significantly. Balanced bidirectional dominance implies a history of selection for intermediate values of the trait. High additive variance indicates that a trait has not been important for fitness, or that selective pressures are inconsistent and that the ability to vary phenotypes rapidly in response to a change in environment is valuable. There is an increasing number of studies in which the genetic architecture of a behavioural characteristic in domesticated strains has been used as the basis for inferences regarding the relation of that trait to fitness. Bruell demonstrated that F_1 hybrids between inbred strains of mice almost always surpassed both parental strains on activity tests.

The poorer performance of te parents was attributed to the inevitable, random loss of some dominant alleles during inbreeding. These losses generally involve different loci in any two strains; thus, in their hybrid offspring the deficiency of one is compensated by the contribution of the other. As might be expected, Bruell found that heterosis was more pronounced in crosses whose parents were distantly related than in crosses between lines with similar origins. More sophisticated modes of genetic analysis were used by Broadhurst and Jinks, who measured open field activity and defecation in six lines of rats and all possible F_1 hybrids. This diallel cross design provides estimates of many genetic parameters.

Their genetic findings were:

- Genes enhancing developmental phenostability tend to be dominant;
- Dominance also occurs for rapid decrease of emotional defecation on successive tests in the open field;
- Exploratory activity is also regulated by dominant genes, but these may have either a negative or a positive sign. From thse observations they concluded that selection has operated to:
- Buffer the fetal and infant rat from external stress;
- Promote a rapid return to normal emotional status after arousal;

- Strike a balance between bold curiosity that could lead to disaster, and timidity that could cause exclusion from essential resources. None of this is surprising, but it is satisfying to find that the genetic findings are consonant with our expectations of what organisms should do in order to prosper.

The genetic approach to reconstruction of evolutionary history is gaining ground, but there is difference of opinion regarding the best way to proceed. For example, Lynch concluded that: "because inbred mice are not weakling, analysis of croses between them may represent a more efficient method for detecting genetic architecture than inbreeding of wild populations." Smith disagrees on the basis of his comparison of the heritabilit of shuttle avoidance in selected wild mice and in inbred strains.

He doubts that it is safe to deduce the selection history of a hypothetical parent popuation from the genetic architecture of contemporary inbred strains of largely unknown origin. Obviously, more research is needed to settle this matter.

Henderson made an effort to escape the bind of first accepting the basic hypothesis of th relation between genetic dominance and fitness of a trait, then conducting a genetic experiment and, from its results deducing the fitness value of that trait. He deduced from observation that a particular trait was related to fitness and then conducted a breeding experiment to determine its genetic architecture. Specifically, he reasoned that mice should be selected for low levels of motor activity during infancy because they are safest when they remain in their nest. If they are accidentally dislodged their chances of being retrieved are greater if they remain quiet and await their mother. He predicted that, in direct contradiction to the situation in adults, low activity would be dominant in infant mice.

A triple test cross confirmed his prediction; the heritability of activity was low and inactivity was dominant. This was attributed by Henderson to a relaxation of selection for infant immobility over many generations of laboratory rearing. In a very different species, Cheng, Shoffner, Phillips, and Shapiro studied imprinting to visual and auditory stimuli in five groups of mallards (Anas platyrhyncs) ranging from a long established game farm line to 25%, 50%, 75%, and 100% wild ancestry. Wild ducklings imprinted better than domesticated ducklings, and heterosis was found in the hybrids, as would be predicted from the hypothesis that traits with adaptive value show little additive variation. Efficient imprinting seemed to be correlated with differences in arousal during early exposure to stimulation.

Confirmations of a prediction do not prove the universal value of a method. However the approach of Henderson and Chang *et al.* is promising and has considerable potential. As Henderon states, the method is most powerful when predictions can be made that a particular kind of behaviour will differ in its contribution to fitness at different parts of its life-span, or in dissimilar environ ments.

The method is obviously limited to species in which multiple distinct lines or strains are available. Principles derived from such studies are, however, likely to have general application. It is only fair to state that the usefulness of this approach to defining adaptive behaviour has been challenged on the basis that the experimental subjects available for analysis of genetic architecture are not representative of natural populations, and that a history of inbreeding distorts the genetic structure typical of a species under natural selection. These warnings must be keptin mind but much of the data obtained in this way seems to substantiate accepted ideas of the characteristics of fitness enhancing behaviour. There is, of course, a possible circularity in such experimens.

One can hypothesize tht high levels of behaviour B_1 are adaptive and thus should have increased fitness during recent evolutionary history. Suppose now that an experiment is performed and low levels of B_1 are found to be inherited in dominant fashion. One could conclude that the original hypothesis was wrong, and that actually low levels of B_1 enhance fitness.

Alternatively one might consider the possibility that B_1 has little to do with fitness but, in the population studied, it is correlated genetically with another behaviour, B_2, that has been selected on the basis of enhancing fitness. The best way out of such circularity is good ethological studies relating B_1 and B_2 independently to survival and propagation.

Another prblem with the attempt to deduce the selective importance of a particular form of behaviour is the choice ofthe method of measuring it. Behaviour geneticists have been sure for years that in an open field C_{57}BL mice were more active than BALB/c mice. Whitford and Sipf found no important differences between these strains when the height of their open field was reduced from 36 inches to 1 inch by insertion of a plxiglass sheet. It is clear that open top, open field activity is very different from contact top, open field activity.

Plant Genetic Diversity

The conservation of plant genetic resources has, in recent years, attracted growing public and scientific interest and political support. There is an increasing awareness of the relevance of biological diversity and its conservation to the health of the biosphere. Particularly urgent is the need to raise agricultural output to meet the basic nutritional needs of increasing populations. The dilemma is that the required increase in food production must be obtained through 'sustainable' forms of agriculture which are less dependent on the use of modern high-yielding varieties bred for intensive, high-input systems. It seems unlikely that increased food production on this scale will come from increasing the area under cultivation. Rather, major solutions will depend upon innovations which exploit sophisticated methods to manipulate plant genomes and the totality of natural genetic variation.

The area of disease resistance will serve to illustrate this argument. Many authors have emphasized the importance of innovation in the area of crop protection since crop losses due to pests and disease may account for between 20 and 40% of productivity worldwide. Indeed, over 50% of the research and development (R&D) spent in the plant breeding/seed industry is focused on the identification and incorporation of resistance traits. In a situation where there is some evidence that the pool of available variation for disease resistance genes within commercial breeding lines is becoming limited (*e.g.* for yellow rust in cereals) the search is on for novel sources of resistance. There are now many examples of biotechnology being put to the service of pest control, through the incorporation into plants of transgenes for toxins, antifeedants, enzyme inhibitors, etc. But, as Woolhouse (1992) has pointed out, such is the long coevolutionary history of pests and pathogens with their hosts that present strategies for interfering with these relationships represent only a small proportion of the likely defence mechanisms used by plants in resisting foreign organisms. This enormous diversity of mechanisms represents an arsenal of ammunition which can be used in the fight to minimize disease and predation.

Are there any hard facts which support these assertions concerning the value of genetic resources to agricultural productivity? Can we put a cost on the loss of diversity? The sources of variation used by the plant breeder range from current breeding lines to wild species and the products of direct gene manipulation. In a recent analysis of the plant breeding/seed industry, Swanson (1996) showed that, over a five-year period, 6.5% of all genetic research within this industry, resulting in a marketed innovation, was concerned with germplasm from wild species and landraces.

This compares with only 2.2% of new germplasm arising from technological approaches of induced mutation. The vast majority of germplasm successfully used by the industry still arises from 'conventional' sources, *i.e.* the heavily exploited commercial cultivars, which at first sight might be taken to suggest that wild resources are relatively unimportant in R&D.

In fact Swanson (1996) argues just the opposite, suggesting that the stock of existing commercial varieties should be seen as the information base from which bioindustries develop innovations, whereas the sources of new diversity (wild species, landraces, induced mutation) should be seen as supplying increments to the information base. In other words, R&D requires an annual injection of 'new' genetic material from natural sources, estimated as amounting to approximately 7% of the stock of material currently within the system. Biodiversity is valuable to industries other than those that seek novel genes for crop plant improvement. Biodiversity has traditionally provided a source of compounds to the pharmaceutical, foodstuffs, crop protection and other industries and here it is possible to give more precise evaluations of its importance. As outlined in Chapter of this volume, drug sales based on natural products from plants were estimated at $US43 billion in 1985 alone and the value of yet undiscovered pharmaceuticals in tropical rainforests is estimated to be as much as $US147 billion to society as a whole. Vast markets exist for the replacement of synthetic biocides, many of which have associated toxicological and environmental problems, by 'natural' compounds, which may have more specific modes of action and fewer side-effects.

Even greater may be the use of genetic resources in more fundamental research where their value in the longer term may be unknown or impossible to estimate accurately at the present time. Diverse genetic material is needed and now regularly utilized to create mapping populations and to study inheritance of a wide range of traits which may or may not be of immediate practical value. Very large numbers of germplasm samples are regularly screened and evaluated for different characteristics so that the underlying molecular biology can be investigated. The intention here is that a greater understanding of fundamental biology will give rise to wealth creation in the future.

Novel Technologies for Biodiversity Assessment and Monitoring

The analysis and characterization of genetic diversity is fundamental to any *ex situ* or *in situ* conservation strategy. Over the years the methods for detecting genetic diversity have expanded from Mendelian analyses of discrete morphological variants, to biometrical approaches, biochemical techniques based upon protein and isozyme profiles, to methods based on DNA sequence variation.

There is now a huge array of molecular marker techniques, and Chapter reviews the major technologies and their appropriate use in addressing key issues in germplasm conservation The application of molecular technologies to increase our understanding of biodiversity at several levels is exemplified by Chapter, which is concerned with the population level, which is concerned with molecular approaches to genetic relatedness. Chapter reviews how the collection of diverse germplasm for *ex situ* purposes and the development of effective *in situ* collections both require an understanding of the spatial and temporal aspects of population structure and associated genetic diversity and the factors that control it.

Studies with low- or single-copy markers are particularly valuable in answering questions about genetic relatedness between species. Chapter illustrates how comparative molecular mapping provides new opportunities for exploring the genetical basis of phenotypic variation within and between plant species. The implication of comparative mapping is that genes controlling specific phenotypic traits will have corresponding map locations in different taxa. This provides a new framework for the identification of homologous genes in germplasm and new tools for the effective utilization of germplasm in crop plant improvement.

This approach is also valuable in the study of quantitative traits controlled by several genes since flanking markers for major genes provide markers for quantitative trait loci (QTL) manipulation and selection and the map-based cloning of at least some genes controlling quantitative traits.

Biotechnological Tools for Conservation

Biotechnology can make a substantial contribution to the quality of germplasm collections and their efficient management. For example, in the management of very large germplasm collections there is a clear requirement for procedures for fast, reliable taxonomic identification of material and the elimination of redundancy. Samples characterized by molecular criteria may prove superior to traditional analyses of phenotype and examples of the use of DNA markers to guide the management of germplasm are explored in Chapter by reference to one of the world's largest collections of germplasm at the International Rice Research Institute.

Not all plant species can be stored in the form of desiccated seed. Socalled 'recalcitrant' species have to be kept in moist, warm conditions and even then have relatively low longevity. In addition, there are many cultivated plants which must be maintained clonally, such as banana, which produce few seeds, or potato, which does not come true from seed.

To these traditional problem species we must now add the products of biotechnology, new categories of germplasm including clones of elite genotypes, and transformed cell lines, for which storage of seed is inappropriate. Biotechnology has impact on storage technologies in two quite different ways. The first of these, concerns the various types of *in vitro* culture which allow the propagation of plant material in an aseptic environment, thus ensuring the production of disease-free stocks and simplifying quarantine procedures for the international exchange of germplasm. The development of sensitive molecular probes to detect low levels of pathogens.

A more radical approach to safe storage and distribution of germplasm, and possibly the only way to significantly reduce the rate of loss in biodiverse tropical forests, within the bounds of available resources, is to develop alternatives to the costly (and inherently vulnerable) methods of *ex situ* storage of seed or tissue cultures, by conserving germplasm in the form of DNA or DNA-rich materials. The latter include those dried specimens conserved in the vast resources of the world's herbaria. The DNA Bank-Net, an international network of DNA repositories, has now been established with both storage and distribution functions. Although technologies for rapid removal, analysis, amplification and secure storage of small amounts of DNA are now well advanced, research is still needed on methods to amplify the entire genomic DNA of a species, which means that storage of DNA or DNA-rich materials should still be considered to be something of an insurance policy rather than a replacement for conventional modes of germplasm storage.

Biotechnology-assisted Crop Improvement

Effective use of genetic resources in plant breeding programmes requires methods to determine whether useful genetic variation exists, and costeffective methods to introduce useful genes into the material. Biotechnology has made a contribution to both these aspects.

The use of genetic maps based on molecular markers linked to genes of interest facilitates the more rapid selection of both simple and complex traits thus accelerating their incorporation into breeding materials. Where traits are controlled by one or a few genes, direct isolation of genes and their incorporation into elite cultivars by increasingly efficient transgenic technologies provide an additional approach to the more traditional methods of plant breeding. This prospect also challenges one of the classical concepts in plant genetic resources work, namely the 'gene pool' concept of Harlan and de Wet (1971), which separated cultivated species and their wild relatives into different categories based upon their 'crossability'.

The value of this concept in guiding the work of the plant breeder must now be re-evaluated in the light of molecular genetic technologies which have not only cast a completely new light on the relatedness of crop plant genomes, but which have effectively widened the potential gene pool of a species to include all other life forms.

What challenges does this imply for the traditional plant genetic resources programmes? Why is the emphasis on conserving species that are related to crops when more exotic and potentially valuable genes may be present in liverworts, or green algae, or other less studied taxa?

Bioprospecting for Novel Genes

The value of genetic resources as a source of novel compounds for various industries has been alluded to above. For centuries biodiversity has provided sources of fuels, medicines, foodstuffs, pesticides, etc. The high returns available from introducing a new drug, for example, plus the realization that biodiversity is rapidly being lost (estimated rates of extinction range from 30 to 300 species per day) has encouraged new attitudes to the exploration and exploitation of biodiversity through 'bioprospecting'. How does biotechnology impact on this traditional area? A fundamental operation in any bioprospecting programme is the screening of thousands of extracts for compounds of interest. Gene technology now allows the speeding up of screening programmes for new compounds through the development of more sophisticated *in vitro* assays. For example, the genes encoding receptor proteins for certain classes of drug, or enzymes that may serve as targets for novel pesticides, may be cloned and expressed on a large scale in high-throughput *in vitro* assays into which thousands of plant extracts may be applied.

Bioprospecting is also concerned with discovering novel genes and natural variants of genes or gene products that may either be used directly or to guide sequence improvement by molecular methods. Chapter gives examples of amino acid sequence improvement in peptide hormones as a result of screening homologous genes from the wild, and this area of biotechnology is likely to develop in parallel with new molecular methods of generating random variation through employing the mutational potential of the polymerase chain reaction (PCR) – *i.e.* 'forced molecular evolution' – or recombinatorial libraries of peptides.

Whilst gene prospecting has provided a new perspective on biodiversity, the area is not without political controversy because these advances tend to favour agriculture and other industries of developed countries, and may actually displace traditional commodity production in resource-rich but underdeveloped countries. Chapter also outlines the development of strategies which integrate conservation and exploitation with economic returns and sustainable development in those countries from which the bioresources were first obtained.

The Exchange of Information and Germplasm

The value of all plant genetic resources lies in their utilization in crop improvement programmes or in other bioindustries. The vast majority of conserved germplasm is available for distribution to the global scientific community but this international distribution poses considerable risks of accidental introduction of non-indigenous plant pathogens and pests. The detection and elimination of such disease-causing agents is therefore essential.

Often, such agents will be present at very low levels and may actually be symptomless. In this context then, new technologies reviewed in Chapter have been of great value in the development of reliable and sensitive nucleic acid- or antibody-based diagnostic probes, and these will be of great importance in optimizing the health and value of germplasm collections. Finally, if the biotechnology revolution has had great impact in developing novel molecular and cellular technologies, no less a revolution has affected radically the way in which the world organizes and disseminates information.

Powerful computers linked by the Internet have facilitated the development of sophisticated and comprehensive informatics systems which enable information on plant genetics and molecular biology to be cross-related to systematic, ecological and other data through international networks. Chapter explains how this information is organized and distributed against a background of continual development.

Plant Genetic Variation

Effective conservation and use of plant genetic resources involve asking many questions about the extent and distribution of genetic variation. Only when the appropriate markers and technologies for describing this variation are accessible can such questions be adequately addressed.

The most appropriate markers for a given question should:

- Be heritable;
- Discriminate between the individuals, populations, or taxa being examined;
- Be easy to measure and evaluate; and
- Provide results that can be compared with results of similar studies.

Molecular markers and assays that meet these criteria were introduced in the 1950s and have proliferated ever since. Use of protein isozymes as genetic markers increased rapidly after their introduction. Nucleic acid markers gained popularity in the 1970s, with the advent of DNA sequencing and restriction fragment analysis. The number of different molecular markers and assays has increased exponentially since the late 1980s, when introduction of the polymerase chain reaction (PCR) made DNA amplification and sequencing possible and affordable.

As the diversity of markers and assays increases, the cost of using them continues to decrease. At the same time, the range of taxa and tissues that can be analysed has expanded. Improved DNA extraction, amplification, and sequencing protocols allow us to analyse ancient DNA, from 13,000-year-old specimens of *Mylodon* (ground sloths) and 17–20 million-year-old *Taxodium* (bald cypress) specimens, and samples as small as single cells or fractions of a seed.

Thus the chances are good that at least one marker assay will be appropriate for any question about variation in any taxa. Furthermore, results of one study can generally be related to results from many other studies. Such opportunities are inevitably accompanied by challenges. None of the challenges are unique to molecular studies, but all become more critical as the number of available marker assays increases. Faced with a bewildering variety of molecular technologies (and their acronyms), choosing a marker and assay that are appropriate for a scientist's particular question, taxa of interest, and practical constraints can be overwhelming. The temptation is great to use techniques that are recommended by coworkers, are frequently used, for which the most information is available, or are the newest – without adequately evaluating the techniques to determine which is most suitable.

Because molecular markers can be measured in virtually all plant taxa and minimal tissue is required, extra care must be taken to verify that the entries in a test array are correctly identified. Since molecular markers are widely applicable and their constraints are not always understood, thoughtful planning is particularly important when designing molecular studies, to ensure that they have an appropriate range of entries, an appropriate number of entries per taxon, and conclusions that stay within the scope of a specific question and test array.

In addition, the vast amount of published research using molecular markers complicates the task of thoughtfully relating one study to others with similar marker assays, research questions, and/or taxa.

Another challenge is formulating research objectives that ask critical questions about plant genetic resources. Descriptive studies that use molecular marker assays to examine variation between and within specific taxa can generate valuable information, but some plant conservation questions are best addressed by experimental studies.

The wide variety of molecular marker assays available provides new ways to address such questions – if the markers are appropriately used. Such challenges can be met successfully and thoughtful decisions can be made if logical frameworks are available to categorize and compare these valuable molecular tools.

A solid foundation for such comparisons has been provided recently. This chapter builds on that foundation by reviewing both early and recently developed molecular markers and assays, citing examples of their use. The types of questions asked about plant genetic resources are described, as are guidelines for selecting appropriate markers and assays to address these questions.

The literature cited here is far from comprehensive, but includes applications of each marker assay in plant genetic research. While the list of molecular techniques will soon be dated, hopefully the framework presented will be helpful for categorizing and comparing new tools as they are made available.

The chapter concludes with a perspective on new marker assays as they apply to, and may change, the study of plant genetic variation.

Types of Questions, Markers, and Marker Assays

All studies of genetic variation have four primary components:

- The research question;
- The marker used to address it;
- The technique used to generate and measure the marker; and
- The method used to analyse the data produced. This chapter focuses on the first three components.

Types of Plant Genetic Resource Questions

Most questions asked about *ex situ* and *in situ* plant genetic variation fit into one of three categories. While boundaries between them are not always clear, these categories are useful when deciding how to address specific questions. Questions of identity are forensic in nature and ask whether two or more samples are genetically the same. These questions concern whether geebank accessions are present in duplicate, whether populations are genetically distinct or are geographical subdivisions of one population, and whether genetic change has occurred in an accession or population over time.

In contrast, questions of location and diagnostics concern the presence and location of a particular allele or nucleotide sequence in a taxon, genebank accession, *in situ* population, individual, chromosome, or cloned DNA segment. These questions are asked to locate populations or individuals that have desirable traits; determine whether traits have been lost from a population, or have been transferred from one population or plant to another; and establish the position of markers on physical or genetic maps.

The differences between diagnostic and forensic questions are significant, but are not always clarified. Both types of question require qualitative information about specific loci. Yet while diagnostic studies often use single-locus markers, each with few alternative states, forensic questions are commonly addressed with information from many multiallelic loci.

Finally, relationship and structure questions examine relatedness or similarity between genotypes, the amount of genetic variation present, and how variation is distributed between individuals, populations, and taxa. Such questions are used to pinpoint gaps in genebank collections, decide which genotypes are high priority for *in situ* conservation, design germplasm sampling strategies and regeneration programmes, construct core collections, study gene flow, and determine optimum sizes for conserved populations or geebank accessions. These issues are generally addressed with information from many loci. Quantitative data can be used, but qualitative data are more appropriate for some phylogenetic and parentage questions.

Types of Molecular Markers

The appropriate markers for a study can discriminate between entries in an array, but are not so polymorphic that important variation is masked by random noise. Molecular markers range from highly conserved to hypervariable, and can be either proteins or nucleic acids. The nucleic acids used as markers include entire genomes, single chromosomes, fragments of DNA or RNA, and single nucleotides.

A wide variety of nucleic acid fragments are utilized as markers. While some occur once in a genome, others are repeated. Many repeated sequences used as markers are noncoding; others are elements of multigene families. Some repeated sequences are interspersed throughout the genome, either distributed randomly or in clusters. These interspersed repeats are common in plant and animal nuclear genomes, and are found in plant (but not animal) mitochondrial genomes. The chloroplast genome contains a large inverted repeat (IR); most angiosperm chloroplasts have two copies, separated by a short single- copy region. Repeat length and (rarely) loss of one copy can vary between taxa.

Much research at present is focused on repeated sequences that occur in tandem. The classes of tandem repeats are distinguished by the length of the core repeat unit, the number of repeat units per locus, and the abundance and distribtion of loci. The names for these classes are themselves varied and have been inconsistently used, but Harding *et al.* (1992) and Tautz (1993) have clarified the nomenclature. Tandem repeats were first reported in the literature as 'satellites' of DNA, detected in CsCl density gradients as fractions with different GC content than the rest of the genome.

These satellites have repeat units that are usually several hundred nucleotides long, with thousands of copies at each of several loci in the nuclear genome. These loci are usually in heterochromatin, often near centromeres. Satellite DNA is present in numerous species. For many satellites, the number of loci and number of repeat units per locus vary between species and higher taxa. Minisatellites (often called variable number of tandem repeat loci, or VNTR loci) are widely used as markers, especially in forensics. The repeat units are usually less than 100 nucleotides long, with tens to hundreds of copies per locus. Thousands of loci in a genome may have similar core repeat units. The number of repeat units at a minisatellite locus can vary greatly between individuals and populations. First described in humans, minisatellites are found in numerous animal species, often near telomeres. They are also common in plants, and are often associated with satellites and centromeres. The number of repeat units per locus is less variable in plants tha in animals, but is still high; plant minisatellites are useful markers for variation between and within species.

As suggested by their name, microsatellites – also called simple sequence repeats (SSRs), or simple sequence length polymorphisms (SSLPs) – have very short repeat units, no more than six nucleotides long. SSRs are more abundant than minisatellites in noncoding regions of the nuclear genome, and are present in some nuclear genes and organelle genomes. The number of repeat units per locus is lower for SSRs than for minisatellites, but can approach 100 in animals and 50 in plants. The abundance and polymorphism of SSRs make them particularly valuable for describin variation between populations and individuals.

Like minisatellites, SSRs were documented first in humans and later in plants. Plants and animals differ in the abundance of specific SSR motifs in the genome. In both plant and animal genomes, chromosomal distribution of SSRs is variable. Some animal SSRs are found near heterochromatin or interspersed repeats, but most are randomly dispersed. Dinucleotide SSRs are randomly distributed in *Arabidopsis thaliana* (L.) Heynh., but other studies have located plant SSRs near genes, highly methylated DNA, satellites, or centromeres.

Tandemly repeated genes are also utilized as markers. Perhaps the most widely used are the nuclear genes that encode ribosomal RNA (rRNA). The three rRNA genes are separated by two internal transcribed spacer regions, generally referred to as ITS1 and ITS2. These genes and spacers form a unit that is tandemly repeated hundreds of times, at one to several loci in the genome.

At each of these loci, the individual repeat units are separated by nontranscribed intergenic spacer (IGS) regions. In the middle region of each IGS are tandem copies of a short subrepeat sequence.

Variation in rRNA gene clusters can be measured at several levels, each evolving at a different rate:

- The number and location of rRNA loci, which is highly conserved;
- The (more variable) number of tandem gene clusters per locus;
- The conserved sequences of the three genes;
- The variable sequences of the ITS regions; and
- The highly variable number of subrepeats in the IGS region.

These features make rRNA gene clusters versatile and informative markers for mapping and phylogenetic analyses.

Types of Molecular Marker Assays

Molecular marker assays are generally classified by whether the molecules evaluated are proteins or nucleic acids, and whether the character analysed in a nucleic acid marker assay is the entire genome, a chromosome, a fragment, or a nucleotide. Alternatively, marker assays can be categorized by the type of character measured. Some methods measure quantitative differences between entries in an array. Others measure qualitative characters, each with two or more possible states. Marker assays also differ in the number of loci evaluated per analysis, whether multiple loci are evaluated simultaneously or sequentially, and the type and amount of information needed about the marker loci before conducting the assay.

Protein Marker Assays: When proteins are used as genetic markers, the assumption is made that any variation between proteins reflects heritable variation in their amino acid sequences. This assumption has long been debated, since protein phenotypes can be affected by factors such as post-translational modification, plant phenology, and environmental conditions during plant growth. For many tasks, however, protein marker assays continue to be useful molecular tools.

Some assays measure immunological properties of proteins. When antigenic proteins of one plant or animal species are injected into an animal (usually a rabbit), antibodies are produced. Genetic differences between entries in an array can be assessed by comparing how each entry's proteins bind to antibody serum (antiserum) from a reference species. This binding affinity is measured by the amount of antibody–antigen precipitate produced or by microcomplement fixation (MCF). Each MCF solution contains antiserum from a reference sample, antigen from a test entry, and a protein complement.

The complement can only bind to antibody that is bound to the antigen. The amount of unfixed complement is measured; this is used to calculate the amount of unbound antigen, which indicates the extent of variation (the immunological distance) between the entry and the reference sample. After MCF assays are conducted for each entry in an array, these immunological distances are compared between entries. MCF has long been used in vertebrate phylogenetic analysis. Until recently, immunological markers were seldom used to study plant variation, perhaps because few suitable proteins had been identified. During the past few years, however, monoclonal antibodies to seed proteins of several cereal species have been produced. Reactivity of these antibodies with antigenic proteins can be evaluated by enzyme-linked immunosorbent assays (ELISAs). Monoclonal antibodies have been used to identify cDNA library clones and describe variation between crop species and cultivars. Other marker assays measure the rate of protein migration through a starch, polyacrylamide, or cellulose acetate gel in response to an electrical current. The migration rate is related to protein size, shape, isoelectric point and/or ionic charge.

In these assays, protein extracts from entries in an array are loaded in adjacent gel lanes and electrophoresed. The gel is then treated with a histochemical stain that reacts with a specific marker protein. Variation between the entries is measured by the positions of their stained proteins on the gel. Many of the proteins used as markers are isozymes – functionally similar forms of an enzyme, encoded by different loci or different alleles at a locus. In general all alleles at a locus are expressed, so allelic isozymes can be analysed as codominant variants.

Isozymes are frequently used to describe variation between and within plant populations, species, and genera; for mapping and diagnostics; and in some cases for identification. When comparing higher taxonomic levels, the probability of genetically different electromorphs migrating to the same gel position is increased. For closely related taxa and populations, however, isozyme analysis is informative and continues to be refined. Immunoelectrophoretic assays measure both protein migration rate in gels and the antibody specificity of proteins. Antigenic proteis are electrophoretically separated, then are detected by their reactions with antibodies. This technique is valuable for diagnostics, mapping, and describing variation between plnt cultivars, species, and higher taxa.

Nucleic Acid Marker Assays: Total Genomic DNA and Chromosome Markers: Although proteins are useful as genetic markers, the phenotypeeach protein describes is determined by the genotype plus the type of tissue sampled, plant phenological stage, environment, and post-translational processing. In addition, the percentage of a genome sampled by protein markers is limited: these markers only describe coding sequences, and not all proteins can be separated and detected by established methods. Some questions are best addressed by markers assumed to be selectively neutral, but this assumption is in question for many proteins. Finally, some proteins are unique to particular taxonomic groups, and thus are suitable as markers only in a limited range of taxa.

In contrast, nucleic acids are present in all organisms and are a common genetic currency for comparing virtually any taxa. Nucleic acid markers describe genotypes, not phenotypes, and can sample both coding and noncoding regions of a genome. DNA methylation and secondary structure can cause artefactual variation in some assays, disrupting the direct relationship between marker and genotype. However, these artefacts can often be detected and eliminated.

The first nucleic acid marker assays measured properties of total genomic DNA. When centrifuged in a CsCl density gradient, DNA separates into bands according to GC content. Because highly repetitive (satellite) sequences differ i GC content from the rest of the genome, these gradients can be used to isolate and quantify satellite DNA.

This method has been utilized to compare satellite DNA content between plant species, and is still used to detect highly repetitive DNA.

Another technique that separates repetitive and low-copy DNA relies on DNA reassociation kinetics. The two complementary strands of a DNA molecule denature when heated and reassociate into duplexes when cooled. At a standard temperature and salt concentration, reassociation rate depends on the percentage of strands with similar sequences. Repetitive sequences reassociate rapidly and at high temperatures, compared to single-copy sequences. Thus highly repetitive, middle repetitive, and single-copy fractions of a genome can be isolated by fragmenting, heating, and cooling double-stranded DNA. Because reassociation rate is affected by fragment length and the pattern of sequence repetition, the type and dispersion of repeats can be measured by varying the lengths of the fragments analysed. This method has been used to characterize and compare repeat sequences between plant species and hybrids.

DNA–DNAhybridization employs the same principles, but with single-copy sequences, to compare entries in an array. When single-stranded DNA from two entries is combined, their overall sequence similarity is related to the temperature at which 50% of the strands form heteroduplexes. This *T*m is compared to the homoduplex *T*m for each entry; D*T*m measures the genetic distance between entries. DNA–DNAhybridization is used to examine variation between species and higher taxa, usually with vertebrates and less often with plants and bacteria. Variation in DNA content of nuclei andchromosomes can be measured by flow cytometry. Particles are isolated, suspended, stained with a DNA-binding fluorochrome, and passed through a cytometer. The fluorescent signal intensity from each particle indicates its DNA content. Flow cytometry was first utilized to analyse mammalian genomes, but has since been used to compare nuclear DNA content and ploidy level between plant families, genera, species, and ecotypes and to evaluate widehybrids. Becaus chrmosomes can be identified by their signal intensity, flow cytometry can be used to generate high-resolution karyotypes and sortchromosomes for mapping.

Karyotypes ofchromosomes prepared in mitotic metaphase squashes have long been used for diagnostics, mapping, and describinggenetic variation between plant species andhybrids. Entries in an array can be characterized by their chromosome number and morphology. In addition, chromosome spread preparations can be treated with stains specific to AT-rich regions, heterochromatin, or nucleolar organizing regions. The stain-specific banding patterns of individua chromosomes are then used as markers to characterize and compare entries.

Chromosome banding is informative for mapping studies and describing variation between geneban accessions and cultivars. Banding is often used in combination with flow cytometry or other techniques. Fo chromosomes that are very small or lack distinctive bands, cytogenetic techniques may have limited use. As high-resolution microscopes andmore sensitive stains are developed, however, the value of these assays continues to increase.

Nucleic Acid Marker Assays:Fragment Markers: Nucleic acid fragments are widely used as genetic markers. As mentioned above, CsCl density gradients and DNA reassociation techniques can detect and roughly quantify satellite DNA sequences. Yet these methos do not measure discrete character differences or variation in the sequences themselves. In addition, differences in GC content between genomes can significantly affect the results. In contrast, fragment markers provide high-resolution qualitative information about sequence variation. Either DNA or RNA can be evaluated, and minimal amounts are required. Numerous fragment marker assays are available; all involve detecting fragments that differ in presence, size, or quantity between entries in an array.

Fragments that are produced, electrophoresed, then detected: In many assays, fragments are produced, separated by agarose or polyacrylamide gel electrophoresis, then detected. The length and conformation of a fragment determine its migration rate in a gel. Several different approaches are taken for detecting fragments. In some assays with purified sequences, all fragments are detected and visualized. The fragments are either labelled before or stained after electrophoresis. Ethidium bromide, silver, fluorescent, and radioactive stains and labels are commonly used. In other assays, specific target fragments are detected and visualized by transfer hybridization. All fragments in an assay are electrophoresed, denatured if double-stranded, then transferred from the gel to a nylon or nitrocellulose membrane. A labelled single-stranded probe complementing the target sequence is hybridized to the membrane and anneals to the target fragments. The extent of mismatch hybridization varies with salt and formamide concentration and the temperature during hybridization. *A variety of probes can be used*: Synthesized oligonucleotides, uncharacterized clones from genomic DNA libraries, fragments isolated from other gels, RNA sequences, or total genomic DNA. The probe may be from the same species as the membrane-bound DNA or from a different species. Until recently most probes were radioactively labelled, but nonradioactive colourimetric, fluorescent, and chemiluminescent labels are increasing in popularit).

Fragments produced by restriction digest: Restriction site analysis is widely used in plant genetics. The principles, protocols and data interpretation have been discussed in detail. Restriction fragments are generated by treating double- stranded DNA with restriction endonucleases (REs), which are enzymes produced by bacteria. Each RE cleaves double-stranded DNA at a specific sequence, usually four to six nucleotides in length. Several hundred REs have been characterized; they vary in recognition sequence and ability to digest methylated DNA. DNA sequence variation between the entries in an array is detected as variation in the number and/or lengths of restriction fragments. This variation can be caused by point mutations at RE recognition sites or by the insertion, deletion or rearrangement of sequences at or between RE sites.

When single- or low-copy fragments are evaluated, variations in fragment length are referred to as restriction fragment length polymorphisms (RFLPs). In some RFLP assays of highly purified DNA sequences, all of the fragments are visualized by labelling them before or staining them after electrophoresis. In most RFLP assays, however, target fragments are detected by hybridized probes. For each entry in an array, the presence or absence of each individual fragment is recorded.

These fragment data are used to infer the presence or absence of each restriction site. For diploid genomes, entries heterozygous for the presence of a restriction site can be distinguished from those that are homozygous. Although few RFLP loci are examined on each gel, every locus can be genetically defined. RFLP analysis can be expensive, but costs per trial can be reduced by probing each membrane several times with different probes. Single- or low-copy RFLPs are used extensively in diagnostics, for mapping, and to verify interspecific hybridization and study genetic relationships and structure. Probes that hybridize across species are especially useful for comparative mapping.

In most restriction site analyses of repetitive DNA, target fragments are detected by probing with a known repeat sequence. The fragments, not the genetically defined sites, are usually the characters analysed. Entries that are homozygous for the presence of a fragment cannot be distinguished from heterozygotes. These assays provide information about variation at many loci simultaneously, but reveal little about the type of sequence variation. Restriction site analysis is often used to study interspersed repeats and satellites. It was utilized in the first assays of hypervariable loci, and is presently the most common method in forensics for evaluating minisatellite VNTR loci. Multilocus restriction fragment profiling of plant genomes is useful for identifying clonal cultivars and studying breeding systems. In some studies, fragment length polymorphism can be interpreted as variation of codominant alleles at a locus. However, because inheritance of markers in multilocus profiles is seldom clearly defined, their use in phylogenetic and gene flow studies may be limited.

Restriction site analysis can generate highly reproducible results, if all steps of the process are carefully controlled. With a variety of probe– enzyme combinations available, assays of the nuclear genome can be used to address many questions about genetic variation at or below the species level. As in isozyme analysis, the probability of genetically different restriction fragments co-migrating on a gel is increased when higher taxonomic levels are compared. Because chloroplast genomes evolve slowly, chloroplast restriction site assays are often used to study variation between higher taxa. In some species, both chloroplast and nuclear genomes are biparentally inherited. In many plant species, however, chloroplasts are either maternally or paternally inherited. For these species, chloroplast restriction site analysis is very informative for studies of gene flow and interspecific hybridization, especially when used in combination with assays of biparentally inherited nuclear markers.

Fragments produced by the polymerase chain reaction: Fragment analysis began in the 1970s with the introduction of RFLPs, and grew exponentially a decade later when the polymerase chain reaction (PCR) was developed. PCR uses the principles of DNA reassociation and the action of DNA polymerase to amplify nucleic acid fragments *in vitro*. Each PCR reaction solution contains a double-stranded DNA template, short single-stranded oligonucleotide primers, thermostable DNA polymerase, enzyme cofactors, and the four deoxynucleotides (dNTPs: dATP, dCTP, dGTP, dTTP).

The template is denatured by heating; the temperature is lowered, and the primers anneal to complementary sequences on the template. The solution is heated again, and the polymerase adds dNTPs to the 3' end of each annealed primer. Double-stranded fragments are produced and serve as templates for the next cycle, which generates fragments with primer sequences at each end. During subsequent cycles, these fragments are amplified at a geometric rate.

With automated thermocyclers and cheaper, improved reagents, PCR is fast becoming more reproducible and affordable. A wide range of known and unknown sequences can be amplified from fresh or ancient DNA and from RNA, and little template is needed; hence PCR is applicable to any taxon. However, contamination is a concern, since contaminant DNA in a reaction solution can function as a template. The specificity and reproducibility of PCR depend on several factors: the concentration and quality of the reaction ingredients; the design and GC content of the primers; and the temperature, duration and number of cycles. However, measures can be taken to minimize and detect contamination and optimize cycling protocols. The disadvantages of PCR are greatly outweighed by the tremendous opportunities it provides. When amplified fragments are used for high-resolution mapping, diagnostics or forensics, they are generally detected by probes. When used to describe variation between and within species, amplified fragments may be detected by labelling primers or dNTPs in the reaction solution, labelling PCR products, or staining gels after electrophoresis.

Numerous PCR-based marker assays have been developed and used in plant genetics. They can be categorized by whether target sequences are known prior to amplification, whether the primer sequences are designed or are arbitrary, the number of primers, and the size range of amplified products. In the first PCR assays the target loci were known single- copy sequences, amplified by two primers designed to complement the regions flanking the loci. Primer pairs have since been designed for numerous single-copy loci in plant and animal nuclear and organelle genomes. These loci are very useful as markers for mapping, diagnostics, phylogenetic analyses, and studying hybridization.

In most genome mapping projects, cloned DNA sequences present at only one site in a genome are used as physical mapping landmarks. Primer pairs have been designed to amplify many of these landmarks as sequence tagged sites (STSs). This strategy is used extensively in the Human Genome Project (HGP) and other mapping programmes. In assays of these designed-primer PCR loci, the character evaluated may be the fragment itself, with presence dominant over absence. Often, however, inheritance can be defined.

For loci in diploid genomes, differences in fragment length can then be analysed as codominant alleles. Both interspersed and tandemly repeated sequences can be amplified by designed-primer PCR. When several interspersed repeat loci are amplified in one sample, each locus is evaluated separately. When tandem repeat loci are amplified, length polymorphism at a locus is assumed to reflect variation in the number of tandem repeat units.

Amplification of human minisatellites with designed primer pairs was soon followed by amplification of plant minisatellites and SSRs. Amplified plant SSRs are informative for mapping, genotyping, diagnostics, and population studies. SSR alleles may differ little in length, and discriminating them can be difficult. New electrophoresis, labelling and imaging technologies are making detection easier.

Development of designed-primer PCR markers involves several steps. Target sequences are identified and isolated, usually from a genomic DNA library; the loci and flanking regions are sequenced; and primer pairs are designed and synthesized. Currently these tasks are expensive, and not all laboratories have the equipment required. Once primers are designed and synthesized, however, the costs for each analysis are not high. The initial expenses of marker development may be reduced by identifying the desired sequences from nucleic acid databases, *e.g.* Genbank or EMBL. However, this is only applicable to species well represented in the databases. As another cost-cutting measure, some primers designed to amplify loci in one species can be used to amplify loci in related taxa. Primers that amplify across taxa are also useful for phylogenetic analyses and studying genome evolution.

For development of designed-primer PCR markers, both target and flanking region sequences must be known. Soon after PCR was introduced, however, its ability to amplify unknown sequences was realised. In inter-repeat PCR, primers that complement interspersed repeats or other known sequences are used to amplify unknown sequences between the repeats. 'Universal' primers have been designed to complement conserved coding sequences in mitochondrial and chloroplast genomes; these primers can be used to amplify the noncoding regions between genes.

Other marker assays require no prior sequence information, for either the target region or the flanking regions. These assays – randomly amplified polymorphic DNA analysis (RAPD), arbitrarily primed PCR (AP-PCR; Welsh and McClelland, 1990), and DNA amplification fingerprinting (DAF) – use one short primer that has an arbitrary sequence. This primer anneals to the template at complementary sequences in both 'sense' and 'antisense' orientation, and the regions between primers in opposite orientation are amplified. The number and length of fragments amplified by one primer can vary; short primers generally produce a large number of small fragments. In single-primer PCR, entries that are homozygous for the presence of a fragment cannot be distinguished from heterozygotes; thus fragment presence is generally dominant over absence. Many loci can be evaluated simultaneously, although little genetic information about each locus is provided. Arbitrary PCR is particularly valuable when no sequence information is available for the taxa being studied. It has been used extensively to examine variation between and within plant species, identify genebank accessions and cultivars, and construct genetic maps.

Many PCR assays are variations or combinations of designed-primer PCR and arbitrary PCR. Most were developed to increase the reproducibility of results and/or to better discriminate between closely related genotypes. In arbitrary PCR, reproducibility and the extent of mismatch primer annealing are highly dependent on the cycling parameters and concentration of reaction components. One solution to this problem is to identify polymorphic fragments in arbitrary PCR assays and then isolate, clone, sequence and design primers for these fragments. This process generates new designed-primer PCR loci but involves many steps. Other assays with increased reproducibility combine features of single-primer and interrepeat PCR: unknown sequences are amplified by one primer that complements an SSR, its flanking region, or both. Finally, 'anchored' PCR uses one designed and one arbitrary primer to amplify unknown sequences.

For genotyping, diagnostics, and assessing variation between close relatives, PCR-based marker assays with high resolution are needed. One such variation is nested PCR. In this technique, one amplification reaction with two designed primers is followed by a second reaction. The first PCR product is the template for the second reaction, which uses two designed primers internal to the first set. Although the possibility of contamination is high, nested PCR is valuable for extremely sensitive genotyping and diagnostics. At present, it is used most often in medical research.

Resolution of arbitrary PCR can be increased by adding stable mini-hairpin sequences to the 5' ends of short primers. This technique greatly improved discrimination between cultivars of centipedegrass [*Eremochloa ophiuroides* (Munro) Hackel]. Other marker assays with improved resolution combine principles of PCR and restriction site analysis. RE digestion of PCR templates before they are amplified was first suggested by Mullis and Faloona (1987). Arbitrary PCR with cleaved templates has since been used to map plant genomes and identify genotypes. Amplified fragment length polymorphism (AFLP) analysis is a variation of cleaved-template PCR. After DNA is digested, a short adaptor sequence is ligated to each end of the restriction fragments. The fragments are then used as templates for single-primer PCR. The primer contains a constant sequence (the adaptor and RE recognition site sequences) plus selective bases that determine the specificity of primer annealing. Both reproducibility and resolution are increased by using AFLPs. Resolution can also be increased by digesting PCR products before electrophoresis. Digested products of both designedprimer PCR and arbitrary PCR have been used to measure variation between genebank accessions, refine RFLP maps, and study interspecific gene flow. Resolution of both PCR and restriction site assays can be improved by denaturing double-stranded DNA fragments before they are electrophoresed. Due to internal point mutations or differences in conformation or GC content, the two single strands from one fragment may have different migration rates on polyacrylamide gels. Both denaturing gradient gel electrophoresis (DGGE) and single-stranded conformation polymorphism (SSCP) analysis can detect differences between closely related genotypes.

Fragments detected directly by probe hybridization. In some marker assays the target sequences are not separated by electrophoresis, but are detected directly by probe hybridization. The types of probes used are the same as in electrophoretic assays, and can be radioactively or nonradioactively labelled. In dot and slot blot hybridization, single-stranded DNA fragments, genomic DNA, or RNA is immobilized on a nylon or nitrocellulose membrane. Labelled probes complementing the target sequence are hybridized to the membrane blot. All loci in the blot are sampled simultaneously; abundance of the target sequence is measured by the intensity of the hybridization signal. Many blots on one membrane can be evaluated at the same time, and the procedure is relatively simple. Because the presence of nontarget sequences in the blot can affect hyridization, precise measurment of target sequence abundance in genomic DNA blots can be difficult. When the blots are carefully prepared and hybridization conditions are controlled, however, these assays can generate valuable information for diagnostic and relationship studies. Dot and slot blots have been used to detect specific alleles of genes and RAPD loci, estimate the copy number of satellites and interspersed repeats, and compare gene expression between organs of a plant.

In another technique, *in situ* hybridization (ISH), probes are hybridized to the same chromosome spread preparations used for karyotypes and banding pattern assays. ISH is used to detect the presence and chromosomal distribution of target sequences. While banding assays are only appropriate with chromosomes that contain specific sequences, ISH can detect a wide range of sequences. Radioactive and nonradioactive probes are used, but the popularity of fluorescent *in situ* hybridization (FISH) is rapidly increasing.

ISH has many applications for genome analysis. These include 'painting' the genome of one species onto chromosomes of another, to study species evolution; painting a genome with one of its own chromosomes, to study chromosome evolution; and characterizing the chromosomal distribution of interspersed repeats. ISH is often used in combination with other marker assays. In wide crosses and somatic hybrids, ISH with species-specific interspersed repeat probes can indicate the abundance and chromosomal locations of transferred sequences; restriction site analysis can provide information about structure and rearrangement of the sequences. Physical mapping with ISH can complement RFLP genetic mapping studies.

Aledo *et al.* used slot blots to estimate the copy number of an interspersed repeat in the maize (*Zea mays* L.)genome, and used FISH to determine the chromosomal distribution of the repeat. With both ISH and heterochromatin-specific staining on the same chromosome preparation, Harrison and Heslop-Harrison (1995) confirmed that two *Brassica* satellite families are located near centromeres. By varying thehybridization stringency, differences between satellite sequences on differentchromosomes were revealed.

ISH can be time consuming, and results can be affected by the specific methods used for chromosome preparation, denaturation, and probehybridization and detection. However, new protocols are increasing the reproducibility of ISH.Fukui *et al.* (1994) used a thermocycler to control temperatures during denaturation, and developed accurate imaging methods for detecting and analysing fluorescent signals. Because of such innovations, the value of ISH as a molecular tool will most likely continue to increase.

Nucleic Acid Marker Assays: Nucleotide Sequencing: Nucleic acid variation can be described most directly by using the nucleotides themselves as discrete characters, each with four possible states. Because nucleotides are the basic units of genetic information in all organisms, sequencing can be used to detect very low levels of variation. Nuclear and organelle genomes of virtually any taxon can be sequenced; minimal amounts of DNA or RNA are needed, and fresh or preserved DNA can be used. With most marker assays, entries in a trial are compared to each other; results are not easily compared across trials that have no standard entries in common. In contrast, sequences from one study can be compared with sequences from any other study. In addition, sequencing can reveal whether variation is due to nucleotide substitution or sequence rearrangement.

Two sequencing methods have been used since the late 1970s, and are described in detail by Hillis *et al.* (1990).

In both methods, labelled fragments are generated from DNA or RNA that has been isolated and purified; the fragments are electrophoresed; and the sequence is 'read' from the fragment patterns. Maxam–Gilbert (chemical) sequencing is used to sequence double-stranded DNA. DNA is endlabelled, then divided into four reaction solutions.

Each solution contains one of four reagents, each of which cleaves DNA at specific bases. The sequence of the DNA determines the lengths of the fragments produced. *In vitro* (dideoxy chain termination) sequencing is applicable to both DNA and RNA. The single-stranded template is divided into four reaction solutions. Each solution also contains the following: a labelled oligonucleotide primer that complements one end of the template; the four dNTPs; and one of four dideoxynucleotides, each of which terminates extension at one of the four bases. In each reaction, the primer is extended along the template. As in chemical sequencing, the template sequence determines the lengths of the fragments produced. Chemical sequencing is preferred for some genomic regions with complex secondary structure; in general, however, the chain-termination method is more commonly used. Since it has been coupled with PCR, combined with fluorescent labelling and automation, this technique is rapidly becoming the method of choice.

Sequencing can be used to address a wide range of plant genetic resource questions about virtually any taxa, but involves considerable investment of money, labour and time. Automation and large-scale production of cheaper reagents have made sequencing more cost effective, but few laboratories can afford to sequence large genome segments from multiple entries. As a result, in many studies only one locus or a few loci are sequenced for a limited test array. While many adjacent nucleotides may be evaluated, together they sample only one locus. Thus sequencing is usually not practical for mapping, multilocus forensics, population genetics, or analyses of intraspecific variation. It is often used for high-resolution diagnostics, studying specific genes, and describing variation between genera and higher taxa. In phylogenetic studies, however, inferring genome evolution from sequence variation at a single locus should be done with caution if at all. Many regions of plant nuclear and organelle genomes have been sequenced and used as markers. These sequences vary greatly in evolutionary rate and the taxonomic levels at which they are employed. Chloroplast genomes evolve slowly; sequences of the entire genome or its *rbc*L gene are informative as markers for describing variation between genera and higher taxa. Nuclear genes that have been sequenced and used as markers include the phytochrome, heat shock protein, and nuclear rRNA gene families. In general, the rRNA coding regions are highly conserved; internal transcribed spacers are more variable, and intergenic spacers are highly polymorphic. Finally, sequencing is often used – in combination with other marker assays – to characterize tandem repeat loci.

While sequencing may be the ultimate molecular tool, caution should be taken to avoid several potential problems. Because each nucleotide is a separate character, sequence fidelity is crucial. Thus nucleic acids must be carefully isolated and purified before they are sequenced. In addition, PCR contamination and amplification errors have more serious effects for sequencing than for fragment analysis. Amplification errors can be minimized by optimizing PCR conditions and using polymerases with proofreading capability. Mistakes during gel reading (manual or automated) must also be considered and minimized. Subjectivity is a real concern when aligning sequences before they are compared. Alignment is complicated and requires assumptions about expected amounts of base substitution and insertion/deletion. Despite these concerns, sequencing remains the tool of choice for many studies.

Guidelines for Selecting Marker Assays

Choosing appropriate marker assays can be challenging, but several considerations can make the task easier.

Important issues are:

- What question is being asked?
- What level of resolution is required?
- How can the results be related to characteristics of the taxa being studied?

Questions of identity are best addressed with high-resolution assays and large numbers of codominant marker loci that are evaluated sequentially. Electrophoretic assays of isozymes and nucleic acid fragments meet most of these criteria. Genetically defined isozymes, amplified SSR loci, and single-copy RFLPs are very informative markers for identification. Due to practical constraints, however, multilocus assays of restriction fragment and PCR markers (VNTRs, RAPDs and AFLPs) are often used. Dot blots, sequencing and monoclonal antibody assays are also appropriate, but the costs of analysing many loci per assay may be prohibitive.

The most appropriate assays for diagnostic questions have high levels of resolution and use genetically defined, codominant markers. Because these questions can often be addressed with one or few loci, useful methods include electrophoretic assays of isozymes, single-copy RFLPs, or SSRs; dot or slot blot assays; and sequencing.

For some questions, monoclonal antibody assays and ISH are also appropriate. Mapping studies often use a combination of tools: isozyme or DNA fragment electrophoresis for genetic maps, and chromosome banding, flow cytometry, ISH, and STS markers for physical mapping.

For questions of relationship and structure, important issues include the type of question asked and the range and levels of taxa evaluated. Because evolutionary questions must be answered with qualitative data, the assays used for diagnostics and identification are appropriate. Questions of similarity can be addressed with these qualitative marker assays or with quantitative assays (microcomplement fixation, flow cytometry, CsCl density gradients, and DNA–DNA hybridization).

The latter two techniques have low levels of resolution, and are generally used to compare genera and higher taxa. Interspecific and intraspecific variation is more appropriately described with isozyme and DNA fragment markers, which have higher levels of resolution. In electrophoretic assays of isozymes and DNA fragments, co-migration of genetically dissimilar electromorphs or fragments is likely when higher taxa are compared. If resolution is too high, however, the assays may be numerically precise but contain random 'evolutionary noise' that masks meaningful variation. Researchers can often decide whether a particular marker assay is suitable for a given study by reviewing the results of previous studies using that technique.

Several characteristics of the taxa being studied are important when selecting a marker and marker assay. One of these characteristics is the mating system. Many apomictic and self-pollinated plant species contain little variation, so techniques with high resolution may be required. When deciding whether to use a chloroplast genome marker, information about chloroplast inheritance is critical.

For many marker assays, polyploidy can be a concern. When polyploid taxa are evaluated, the number of loci sampled, the problem of dissimilar fragments co-migrating, and the level of data complexity are often increased. This complexity may be reduced by using genetically defined isozymes, chloroplast genes, genes that evolve slowly, or single-copy markers – rather than RAPDs or multilocus VNTR profiles.

Practical considerations are often the most important. The amount and quality of proteins or nucleic acids required can differ greatly between marker assays. Sequencing and PCR-based assays require little template and can be used to analyse ancient DNA samples, but are very sensitive to contamination. DNA–DNA hybridization requires large amounts of DNA, but small amounts of contaminant DNA are less of a problem. For some species, fresh tissue is required for isozyme electrophoresis. For some species and marker techniques, the presence of secondary compounds can complicate storage and extraction of DNA or protein, and can interfere with marker detection.

Marker assays also vary in the amount of setup work required and the extent of sequence information needed before conducting the assay. Karyotyping, ISH, and RFLP assays involve preparing chromosome spreads and/or probes. In contrast, some PCR-based assays are streamlined for high throughput and require little setup work. Sequence information is needed for probe hybridization and designed-primer PCR assays, but not for arbitrary PCR. Finally, the resources available must be considered. Minimal equipment is needed for protein electrophoresis, but other techniques require thermocyclers, fluorescent microscopes, cytometers and radioactive labelling.

Choices between newly developed and well-documented techniques should be made in light of the expertise and personnel available. Other considerations are reproducibility, number of entries and loci sampled per analysis, and cost per unit of information. Comparative studies can make selection of a marker assay easier and reveal where costs can be reduced. Costs and technologies are constantly changing, however, and the comparisons themselves may be expensive and time consuming; their value must be weighed against the resources they require and the necessity of continual updates. When comparative studies are not feasible, thoughtful consideration of the issues described here may be helpful.

Future Dimensions of Genetic Analysis

The goals and expectations for analysing plant genetic variation parallel those established across many other fields of biological research, from agriculture, ecology, and evolution to the medical sciences. In all of these fields, future genetic marker assays must incorporate methods to detect, describe, interpret, and store DNA sequence information. Molecular tools of the future are expected to be user friendly, accurate, precise, high throughput, low cost, and potentially automated.

DNA sequence information is the foundation for developing and applying genetic markers to questions of biological variation, whether *in situ* or *ex situ*. Researchers who develop and use sequence-based marker assays for diagnostic, forensic and relationship studies will continue to benefit greatly from information and technologies generated by the international Human Genome Project (HGP). Since its beginning, the HGP has endeavoured to develop genetic and physical maps and determine DNA sequences for the genomes of humans and several model organisms. As part of this effort, genetic analysis methods have been improved significantly. Notable achievements – now taken for granted by many – include new types of genetic markers (particularly those based on PCR), more efficient cloning vectors, the introduction of STS markers, and improved technology and automation for DNA sequencing. Underlying this progress is recognition that the successful development of new technologies for genetic research has been, and will continue to be, critical to many future scientific breakthroughs.

In addition to the HGP, other complementary projects also support initiatives for the development of new molecular marker assays. One such project is the Advanced Technology Programme (ATP) of the US National Institute of Standards and Technology. Among the ATP's goals is working with industry to deliver DNA diagnostic tools to a variety of users, at onetenth to one-hundredth of the current price. Another goal is helping industry make similar reductions in DNA sequencing costs and make DNA sequencing apparatus available at about one-third of the current price. Many biological questions once considered recalcitrant due to the number and cost of the assays required will be reconsidered in the light of new technological developments. The following section includes examples of progress that can be readily applied to the description of plant genetic variation. Also noted are limitations that are unique to plant-based studies, and are not presently addressed by most genetic analysis programmes.

Technology Transfer from the Human Genome Project

Technology transfer from the Human Genome Project (HGP) to the study of plant genetic resources has been very important in two areas:

- Sequence characterization and marker development; and
- Integrating all stages in genetic analysis projects, from template preparation through data interpretation.

Some of these innovations are now being used in plant genetic research; others will require some level of optimization before they are applied.

Plant genetic variation is best characterized with a large number of highly informative, easily analysed molecular markers. While the number of PCR-based markers will most likely continue to increase, the question at hand is how these markers will be detected. Manual, autoradiographic detection of amplified markers presently predominates in plant genetic research. However, ongoing programmes in human genotyping via semiautomated, fluorescent detection of SSRs are progressing at an incredible pace. Reed *et al.* (1994) reported that semi-automated technology can genotype over 100 000 samples, using approximately 250 marker loci, in less than six months. Moreover, Schwengel *et al.* (1994) found that data generated by this method were at least as accurate, more efficient, less labour intensive, and as cost effective as data generated by standard radiolabelling techniques.

Human genotyping kits are available for approximately 400 markers that define a ±10 cM resolution genomic index map. Kits for plant species and families are in development. Although cost remains an issue, these marker assays are clearly capable of handling the large arrays commonly encountered in plant conservation and breeding programmes. The costs of generating and detecting numerous marker loci for each of many individuals can be reduced by evaluating several of an individual's loci in a single lane on a gel. In the studies cited above, SSR loci from an individual were multiplexed at the electrophoresis level: several loci were amplified in separate reactions, the PCR products were pooled, and the pooled sample was electrophoresed. In contrast, several laboratories are developing methods to multiplex at the reaction level, amplifying several SSR loci in one tube. Eleven putative loci in canola (*Brassica napus* L.) have been amplified simultaneously, using 11 pairs of fluorescently labelled primers in one reaction. The loci were distinguished from each other by fragment size and fluorescent label colour. This approach is also under way with SSRs of maize and sorghum [*Sorghum bicolour* (L.) Moench] Another molecular marker used in human genetic research reveals discrete changes (single nucleotide substitutions) in a specific DNA sequence. Variation of these polymorphic sequence tagged sites (pSTS) is the most frequent and widely distributed type of sequence variation in the genome. These individual nucleotide markers are usually biallelic within a population, and are generally less polymorphic than SSRs and other tandem repeats. However, a number of closely linked pSTSs can be combined into a multilocus marker that is as informative as an SSR.

Developing and assaying pSTS markers involve several steps. Random genomic DNA clones (each 300–600 bp long) from selected plant species are sequenced, then single nucleotide differences between individuals are identified by comparing automated sequencing electrophoregrams. Biallelic pSTSs are selected, and primer pairs are designed to amplify them. After a pSTS locus is amplified, the oligonucleotide ligation assay (OLA) is used to label the PCR products and discriminate the two allelic forms. In this assay, the PCR products are denatured and then serve as templates for a ligation detection reaction (LDR). This reaction uses three 'probes': two diagnostic probes, each of which complements one of the two pSTS alleles, and one labelled probe that is common to both alleles.

The diagnostic probes anneal to complementary template sequences, then are ligated to the labelled probe. The two diagnostic probes differ in length. Thus when the LDR-labelled fragments are electrophoresed, pSTS allelic variation between entries can be detected as variation in migration rate of fragments on the gel, due to differences in fragment length. As in many PCR-based marker assays, significant costs are presently incurred for pSTS marker development. Once markers are developed, however, the costs for each assay are not high. The previously highlighted markers are usually associated with noncoding regions of the genome. Complementary efforts are under way to develop high-throughput, automated methods to partially sequence cDNA clones, then use them to identify expressed genes in many organisms, including plants. Thanks to the rapid proliferation of expressed sequence tag (EST) databases, at present the probable gene function of a cDNA clone can often be inferred solely from the similarity of nucleotide or amino acid sequences between the clone and genes of known function. Using these techniques to compare isoforms of a gene (particularly its untranslated 3' regions) between members of a plant species or family, PCR-based markers that describe variation of this gene could be developed. When assays using these single-gene markers are compared with multilocus studies of the taxon, a more complete picture of genetic variation may be revealed.

Present Limitations on Describing Plant Genetic Variation with Molecular Tools

Human genetic research often focuses on genomic mapping and analysing specific genes. In contrast, many plant genetic resource questions are best addressed by evaluating numerous individuals from many taxa. Thus widespread use of molecular markers in plant genetic research will involve preparing and analysing large amounts of DNA template and assaying many individuals at the same time. However, techniques have not yet been developed to extract, clone, replicate or amplify DNA in nanolitre volumes. Also needed are techniques for pooling DNA samples from individuals in a population, assaying the pooled sample, and generating a comprehensive picture of population variation. In the future, sequence-based approaches may be useful for these tasks. At present, high costs make these approaches impractical for many high-throughput applications.

Chapter 7

Chromosomes

Introduction

A chromosome is an organized building of DNA and protein that is found in cells. It is a single piece of coiled DNA containing many genes, regulatory elements and other nucleotide sequences. Chromosomes also contain DNA-bound proteins, which serve to package the DNA and control its functions.

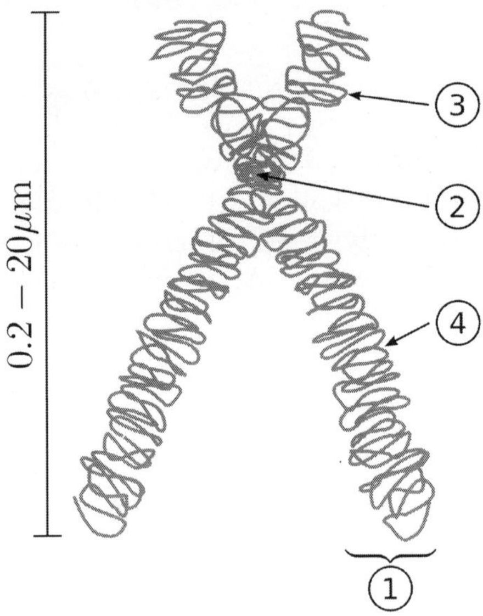

Figure 1: Diagram of a replicated and condensed metaphase eukaryotic chromosome. (1) Chromatid – one of the two identical parts of the chromosome after S phase. (2) Centromere – the point where the two chromatids touch, and where the microtubules attach. (3) Short arm. (4) Long arm.

Chromosomes vary widely between different organisms. The DNA molecule may be circular or linear, and can be composed of 10,000 to 1,000,000,000 nucleotides in a long chain. Typically eukaryotic cells

(cells with nuclei) have large linear chromosomes and prokaryotic cells (cells without defined nuclei) have smaller circular chromosomes, although there are many exceptions to this rule. Furthermore, cells may contain more than one type of chromosome; for example, mitochondria in most eukaryotes and chloroplasts in plants have their own small chromosomes.

In eukaryotes, nuclear chromosomes are packaged by proteins into a condensed structure called chromatin. This allows the very long DNA molecules to fit into the cell nucleus. The structure of chromosomes and chromatin varies through the cell cycle. Chromosomes are the essential unit for cellular division and must be replicated, divided, and passed successfully to their daughter cells so as to ensure the genetic diversity and survival of their progeny. Chromosomes may exist as either duplicated or unduplicated—unduplicated chromosomes are single linear strands, whereas duplicated chromosomes (copied during synthesis phase) contain two copies joined by a centromere. Compaction of the duplicated chromosomes during mitosis and meiosis results in the classic four-arm structure (pictured to the right). Chromosomal recombination plays a vital role in genetic diversity. If these structures are manipulated incorrectly, through processes known as chromosomal instability and translocation, the cell may undergo mitotic catastrophe and die, or it may aberrantly evade apoptosis leading to the progression of cancer.

In practice "chromosome" is a rather loosely defined term. In prokaryotes and viruses, the term genophore is more appropriate when no chromatin is present. However, a large body of work uses the term chromosome regardless of chromatin content. In prokaryotes DNA is usually arranged as a circle, which is tightly coiled in on itself, sometimes accompanied by one or more smaller, circular DNA molecules called plasmids. These small circular genomes are also found in mitochondria and chloroplasts, reflecting their bacterial origins. The simplest genophores are found in viruses: these DNA or RNA molecules are short linear or circular genophores that often lack structural proteins.

History

Chromosomes as Vectors of Heredity: In a series of experiments, Theodor Boveri gave the definitive demonstration that chromosomes are the vectors of heredity. His two principles were based upon the *continuity* of chromosomes and the *individuality* of chromosomes.

It is the second of these principles that was so original. Boveri was able to test the proposal put forward by Wilhelm Roux, that each chromosome carries a different genetic load, and showed that Roux was right. Upon the rediscovery of Mendel, Boveri was able to point out the connection between the rules of inheritance and the behaviour of the chromosomes. It is interesting to see that Boveri influenced two generations of American cytologists: Edmund Beecher Wilson, Walter Sutton and Theophilus Painter were all influenced by Boveri (Wilson and Painter actually worked with him). In his famous textbook *The Cell,* Wilson linked Boveri and Sutton together by the Boveri-Sutton theory. Mayr remarks that the theory was hotly contested by some famous geneticists: William Bateson, Wilhelm Johannsen, Richard Goldschmidt and T.H. Morgan, all of a rather dogmatic turn-of-mind. Eventually complete proof came from chromosome maps in Morgan's own lab.

Chromosomes in Eukaryotes

Eukaryotes (cells with nuclei such as those found in plants, yeast, and animals) possess multiple large linear chromosomes contained in the cell's nucleus. Each chromosome has one centromere, with one or two arms projecting from the centromere, although, under most circumstances, these arms are not visible as such. In addition, most eukaryotes have a small circular mitochondrial genome, and some eukaryotes may have additional small circular or linear cytoplasmic chromosomes. In the nuclear chromosomes of eukaryotes, the uncondensed DNA exists in a semi-ordered structure, where it is wrapped around histones (structural proteins), forming a composite material called chromatin.

Eukaryotic Chromosome Fine Structure

Eukaryotic chromosome fine structure refers to the structure of sequences for eukaryotic chromosomes. Some fine sequences are included in more than one class, so the classification listed is not intended to be completely separate.

Chromosomal Characteristics

Some sequences are required for a properly functioning chromosome:

- *Centromere:* Used during cell division as the attachment point for the spindle fibres.

- *Telomere:* Used to maintain chromosomal integrity by capping off the ends of the linear chromosomes. This region is a microsatellite, but its function is more specific than a simple tandem repeat.

Structural Sequences

Other sequences are used in replication or during interphase with the physical structure of the chromosome.

- Ori, or Origin: Origins of replication.
- MAR: Matrix attachment regions, where the DNA attaches to the nuclear matrix.

Protein-coding Genes

Regions of the genome with protein-coding genes include several elements:

- Enhancer regions (normally up to a few thousand basepairs upstream of transcription).
- Promoter regions (normally less than a couple of hundred basepairs upstream of transcription) include elements such as the TATA and CAAT boxes, GC elements, an initiator, *etc.*
- Exons are the part of the transcript that will eventually be transported to the cytoplasm for translation. When discussing gene with alternate splicing, an exon is a portion of the transcript that could be translated, given the correct splicing conditions. The exons can be divided into three parts
  - The coding region is the portion of the mRNA that will eventually be translated.
  - Upstream untranslated region (5' UTR) can serve several functions, including mRNA transport, and initiation of translation (including, portions of the Kozak sequence). They are never translated into the protein (excepting various mutations).
  - The 3' region downstream from the stop codon is separated into two parts:
    - 3' UTR is never translated, but serves to add mRNA stability. It is also the attachment site for the poly-A tail. The poly-A tail is used in the initiation of translation and also seems to have an effect on the long-term stability (aging) of the mRNA.

- An unnamed region after the poly-A tail, but before the actual site for transcription termination, is spliced off during transcription, and so does not become part of the 3' UTR. Its function, if any, is unknown.
- Introns are intervening sequences between the exons that are never translated. Some sequences inside introns function as miRNA, and there are even some cases of small genes residing completely within the intron of a large gene. For some genes (such as the antibody genes), internal control regions are found inside introns. These situations, however, are treated as exceptions.

Genes that are used as RNA

Many regions of the DNA are transcribed with RNA as the functional form:

- *rRNA:* Ribosomal RNA are used in the ribosome.
- tRNA: Transfer RNA are used in the translation process by bringing amino acids to the ribosome.
- *snRNA:* Small nuclear RNA are used in spliceosomes to help the processing of pre-mRNA.
- *gRNA:* Guide RNA are used in RNA editing.
- miRNA: Micro RNA are small (approximately 24 nucleotides) that are used in gene silencing.
- *snoRNA:* Small nucleolar RNA are used to help process and construct the ribosome.

Other RNAs are transcribed and not translated, but have undiscovered functions.

Repeated Sequences

Repeated sequences are of two basic types: unique sequences that are repeated in one area; and repeated sequences that are interspersed throughout the genome.

Satellites

Satellites are unique sequences that are repeated in tandem in one area. Depending on the length of the repeat, they are classified as either:

- *Minisatellite:* Short repeats of nucleotides.
- *Microsatellite:* Very short repeats of nucleotides. Some trinucleotide repeats are found in coding regions. Most are

found in noncoding regions. Their function is unknown, if they have any specific function. They are used as molecular markers and in DNA fingerprinting.

Interspersed Sequences

Interspersed sequences are tandem repeats, with sequences that are found interspersed across the genome. They can be classified based on the length of the repeat as:

- *SINE:* Short interspersed sequences. The repeats are normally a few hundred base pairs in length. These sequences constitute about 13% of the human genome with the specific *Alu* sequence accounting for 5%.
- *LINE:* Long interspersed sequences. The repeats are normally several thousand base pairs in length. These sequences constitute about 21% of the human genome.

Both of these types are classified as retrotransposons.

Retrotransposons

Retrotransposons are sequences in the DNA that are the result of retrotransposition of RNA. LINEs and SINEs are examples where the sequences are repeats, but there are non-repeated sequences that can also be retrotransposons.

Other Sequences

Typical eukaryotic chromosomes contain much more DNA than is classified in the categories above. The DNA may be used as spacing, or have other as-yet-unknown function. Or, they may simply be random sequences of no consequence.

Chromatin

Chromatin is the combination of DNA and proteins that makes up chromosomes. It is found inside the nuclei of eukaryotic cells. It is divided between heterochromatin (condensed) and euchromatin (extended) forms. The major components of chromatin are DNA and histone proteins, although other proteins have prominent roles too. The functions of chromatin are to package DNA into a smaller volume to fit in the cell, to strengthen the DNA to allow mitosis and meiosis, and to serve as a mechanism to control expression and DNA replication. Chromatin contains genetic material-instructions to direct cell functions. Changes in chromatin structure are affected by chemical modifications of histone proteins such as methylation (of DNA and proteins) and acetylation (of proteins), and by non-histone, DNA-binding proteins.

Basic Structure

In simple terms, there are three levels of chromatin organization:

1. DNA wraps around histone proteins forming nucleosomes-the "beads on a string" structure.
2. A 30 nm condensed chromatin fibre consisting of nucleosome arrays in their most compact form.
3. Higher-level DNA packaging into the metaphase chromosome.

These structures do not occur in prokaryotic cells. Examples of cells with more extreme packaging are spermatozoa and avian red blood cells.

During spermiogenesis, the spermatid's chromatin is remodelled into a more spaced-packaged, widened, almost crystal-like structure. This process is associated with the cessation of transcription and involves nuclear protein exchange. The histones are mostly displaced, and replaced by protamines (small, arginine-rich proteins). It should also be noted that, during mitosis, while most of the chromatin is tightly compacted, there are small regions that are not as tightly compacted. These regions often correspond to promoter regions of genes that were active in that cell type prior to entry into cromitosis. The lack of compaction of these regions is called bookmarking, which is an epigenetic mechanism believed to be important for transmitting to daughter cells the "memory" of which genes were active prior to entry into mitosis. This bookmarking mechanism is needed to help transmit this memory because transcription ceases during mitosis.

During Interphase

The structure of chromatin during interphase is optimised to allow easy access of transcription and DNA repair factors to the DNA while compacting the DNA into the nucleus. The structure varies depending on the access required to the DNA. Genes that require regular access by RNA polymerase require the looser structure provided by euchromatin.

Change in Structure

Chromatin undergoes various forms of change in its structure. Histone proteins, the foundation blocks of chromatin, are modified by various post-translational modification to alter DNA packing.

Acetylation results in the loosening of chromatin and lends itself to replication and transcription. When certain residues are methylated they hold DNA together strongly and restrict access to various enzymes. A recent study showed that there is a bivalent structure present in the chromatin: methylated lysine residues at location 4 and 27 on histone 3. It is thought that this may be involved in development; there is more methylation of lysine 27 in embryonic cells than in differentiated cells, whereas lysine 4 methylation positively regulates transcription by recruiting nucleosome remodelling enzymes and histone acetylases.

Polycomb-group proteins play a role in regulating genes through modulation of chromatin structure.

DNA Structure

The vast majority of DNA within the cell is the normal DNA structure. However in nature DNA can form three structures, A-, B- and Z-DNA. A and B chromosomes are very similar, forming right-handed helices, while Z-DNA is a more unusual left-handed helix with a zig-zag phosphate backbone. Z-DNA is thought to play a specific role in chromatin structure and transcription because of the properties of the junction between B-and Z-DNA.

At the junction of B-and Z-DNA one pair of bases is flipped out from normal bonding. These play a dual role of a site of recognition by many proteins and as a sink for torsional stress from RNA polymerase or nucleosome binding.

The Nucleosome and "Beads-on-a-string"

The basic repeat element of chromatin is the nucleosome, interconnected by sections of linker DNA, a far shorter arrangement than pure DNA in solution.

In addition to the core histones, there is the linker histone, H1, which contacts the exit/entry of the DNA strand on the nucleosome. The nucleosome core particle, together with histone H1, is known as a chromatosome. Nucleosomes, with about 20 to 60 base pairs of linker DNA, can form, under non-physiological conditions, an approximately 10 nm "beads-on-a-string" fibre.

The nucleosomes bind DNA non-specifically, as required by their function in general DNA packaging. There are, however, large DNA sequence preferences that govern nucleosome positioning. This is due primarily to the varying physical properties of different DNA sequences:

For instance, adenosine and thymine are more favourably compressed into the inner minor grooves. This means nucleosomes can bind preferentially at one position approximately every 10 base pairs (the helical repeat of DNA)-where the DNA is rotated to maximise the number of A and T bases that will lie in the inner minor groove.

30 nm Chromatin Fibre

With addition of H1, the "beads-on-a-string" structure in turn coils into a 30 nm diameter helical structure known as the 30 nm fibre or filament. The precise structure of the chromatin fibre in the cell is not known in detail, and there is still some debate over this.

This level of chromatin structure is thought to be the form of euchromatin, which contains actively transcribed genes. EM studies have demonstrated that the 30 nm fibre is highly dynamic such that it unfolds into a 10 nm fibre ("beads-on-a-string") structure when transversed by an RNA polymerase engaged in transcription.

The existing models commonly accept that the nucleosomes lie perpendicular to the axis of the fibre, with linker histones arranged internally. A stable 30 nm fibre relies on the regular positioning of nucleosomes along DNA. Linker DNA is relatively resistant to bending and rotation. This makes the length of linker DNA critical to the stability of the fibre, requiring nucleosomes to be separated by lengths that permit rotation and folding into the required orientation without excessive stress to the DNA. In this view, different length of the linker DNA should produce different folding topologies of the chromatin fibre. Recent theoretical work, based on electron-microscopy images of reconstituted fibres support this view.

Spatial Organization of Chromatin in the Cell Nucleus

The layout of the genome within the nucleus is not random-specific regions of the genome have a tendency to be found in certain spaces. Specific regions of the chromatin are enriched at the nuclear membrane, while other regions are bound together by protein complexes. The layout of this is not, however, well characterised apart from the compaction of one of the two X chromosomes in mammalian females into the Barr body.

This serves the role of permanently deactivating these genes, which prevents females getting a 'double dose' relative to males. The extent to which the inactive X is actually compacted is a matter of some controversy.

Chromatin and Bursts of Transcription

Fluctuations between open and closed chromatin may contribute discontinuity of transcription, or transcriptional bursting. Other factors are probably involved, such as the association and dissociation of transcription factor complexes with chromatin. The phenomenon, as opposed to simple probabilistic models of transcription, can account for the high variability in gene expression occurring between cells in isogenic populations.

Metaphase Chromatin

The metaphase structure of chromatin differs vastly to that of interphase. It is optimised for physical strength and manageability, forming the classic chromosome structure seen in karyotypes. The structure of the condensed chromosome is thought to be loops of 30 nm fibre to a central scaffold of proteins. It is, however, not well characterised.

The physical strength of chromatin is vital for this stage of division to prevent shear damage to the DNA as the daughter chromosomes are separated. To maximise strength the composition of the chromatin changes as it approaches the centromere, primarily through alternative histone H1 anologues.

Non-histone Chromosomal Proteins

The proteins that are found associated with isolated chromatin fall into several functional categories:

- chromatin-bound enzymes
- high mobility group proteins (HMG)
- transcription factors
- scaffold proteins
- transition proteins (testis specific proteins)
- protamines (present in mature sperm).

Proteins associated with chromatin are those involved in DNA transcription, replication and repair, and in post-translational modification of histones. They include various types of nucleases and proteases. Scaffold proteins encompass chromatin proteins such as *insulators*, domain *boundary factors* and *cellular memory modules* (CMMs).

Chromatin: Alternative Definitions

1. Simple and concise definition: Chromatin is DNA plus the proteins (and RNA) that package DNA within the cell nucleus.

2. A biochemists' operational definition: Chromatin is the DNA/protein/RNA complex extracted from eukaryotic lysed interphase nuclei. Just which of the multitudinous substances present in a nucleus will constitute a part of the extracted material will depend in part on the technique each researcher uses. Furthermore, the composition and properties of chromatin vary from one cell type to the another, during development of a specific cell type, and at different stages in the cell cycle.
3. The *DNA + histone = chromatin* definition: The DNA double helix in the cell nucleus is packaged by special proteins termed histones. The formed protein/DNA complex is called chromatin. The structural entity of chromatin is the nucleosome.

Chromatin is the complex of DNA and protein found in the eukaryotic nucleus, which packages chromosomes. The structure of chromatin varies significantly between different stages of the cell cycle, according to the requirements of the DNA.

Interphase Chromatin

During interphase (the period of the cell cycle where the cell is not dividing), two types of chromatin can be distinguished:

- Euchromatin, which consists of DNA that is active, e.g., being expressed as protein.
- Heterochromatin, which consists of mostly inactive DNA. It seems to serve structural purposes during the chromosomal stages. Heterochromatin can be further distinguished into two types:
  - *Constitutive heterochromatin*, which is never expressed. It is located around the centromere and usually contains repetitive sequences.
  - *Facultative heterochromatin*, which is sometimes expressed.

Individual chromosomes cannot be distinguished at this stage – they appear in the nucleus as a homogeneous tangled mix of DNA and protein.

Metaphase Chromatin and Division

In the early stages of mitosis or meiosis (cell division), the chromatin strands become more and more condensed. They cease to function as accessible genetic material (transcription stops) and become a compact transportable form. This compact form makes the individual chromosomes visible, and they form the classic four arm structure, a pair of sister chromatids attached to each other at the centromere. The shorter arms are called *p arms* (from the French *petit*, small) and the longer arms are called *q arms* (*q* follows *p* in the Latin alphabet). This is the only natural context in which individual chromosomes are visible with an optical microscope.

During divisions, long microtubules attach to the centromere and the two opposite ends of the cell. The microtubules then pull the chromatids apart, so that each daughter cell inherits one set of chromatids. Once the cells have divided, the chromatids are uncoiled and can function again as chromatin. In spite of their appearance, chromosomes are structurally highly condensed, which enables these giant DNA structures to be contained within a cell nucleus.

The self-assembled microtubules form the spindle, which attaches to chromosomes at specialized structures called kinetochores, one of which is present on each sister chromatid. A special DNA base sequence in the region of the kinetochores provides, along with special proteins, longer-lasting attachment in this region.

Chromosomes in Prokaryotes

The prokaryotes – bacteria and archaea – typically have a single circular chromosome, but many variations do exist. Most bacteria have a single circular chromosome that can range in size from only 160,000 base pairs in the endosymbiotic bacterium *Candidatus Carsonella ruddii*, to 12,200,000 base pairs in the soil-dwelling bacterium *Sorangium cellulosum*. Spirochaetes of the genus *Borrelia* are a notable exception to this arrangement, with bacteria such as *Borrelia burgdorferi*, the cause of Lyme disease, containing a single linear chromosome.

Structure in Sequences

Prokaryotic chromosomes have less sequence-based structure than eukaryotes. Bacteria typically have a single point (the origin of replication) from which replication starts, whereas some archaea contain multiple replication origins. The genes in prokaryotes are often organized in operons, and do not usually contain introns, unlike eukaryotes.

DNA Packaging

Prokaryotes do not possess nuclei. Instead, their DNA is organized into a structure called the nucleoid. The nucleoid is a distinct structure and occupies a defined region of the bacterial cell. This structure is, however, dynamic and is maintained and remodeled by the actions of a range of histone-like proteins, which associate with the bacterial chromosome. In archaea, the DNA in chromosomes is even more organized, with the DNA packaged within structures similar to eukaryotic nucleosomes.

Bacterial chromosomes tend to be tethered to the plasma membrane of the bacteria. In molecular biology application, this allows for its isolation from plasmid DNA by centrifugation of lysed bacteria and pelleting of the membranes (and the attached DNA). Prokaryotic chromosomes and plasmids are, like eukaryotic DNA, generally supercoiled. The DNA must first be released into its relaxed state for access for transcription, regulation, and replication.

Number of Chromosomes in Various Organisms

Eukaryotes

These tables give the total number of chromosomes (including sex chromosomes) in a cell nucleus. For example, human cells are diploid and have 22 different types of autosome, each present as two copies, and two sex chromosomes. This gives 46 chromosomes in total. Other organisms have more than two copies of their chromosomes, such as bread wheat, which is *hexaploid* and has six copies of seven different chromosomes – 42 chromosomes in total.

Normal members of a particular eukaryotic species all have the same number of nuclear chromosomes. Other eukaryotic chromosomes, i.e., mitochondrial and plasmid-like small chromosomes, are much more variable in number, and there may be thousands of copies per cell. Asexually reproducing species have one set of chromosomes, which are the same in all body cells. However, asexual species can be either haploid or diploid.

Sexually reproducing species have somatic cells (body cells), which are diploid [2n] having two sets of chromosomes, one from the mother and one from the father. Gametes, reproductive cells, are haploid [n]: They have one set of chromosomes. Gametes are produced by meiosis of a diploid germ line cell. During meiosis, the matching chromosomes of father and mother can exchange small parts of themselves (crossover), and thus create new chromosomes that are not inherited solely from either parent. When a male and a female gamete merge (fertilization), a new diploid organism is formed.

Some animal and plant species are polyploid [Xn]: They have more than two sets of homologous chromosomes. Plants important in agriculture such as tobacco or wheat are often polyploid, compared to their ancestral species. Wheat has a haploid number of seven chromosomes, still seen in some cultivars as well as the wild progenitors. The more-common pasta and bread wheats are polyploid, having 28 (tetraploid) and 42 (hexaploid) chromosomes, compared to the 14 (diploid) chromosomes in the wild wheat.

Prokaryotes

Prokaryote species generally have one copy of each major chromosome, but most cells can easily survive with multiple copies. For example, *Buchnera*, a symbiont of aphids has multiple copies of its chromosome, ranging from 10–400 copies per cell. However, in some large bacteria, such as *Epulopiscium fishelsoni* up to 100,000 copies of the chromosome can be present. Plasmids and plasmid-like small chromosomes are, as in eukaryotes, very variable in copy number. The number of plasmids in the cell is almost entirely determined by the rate of division of the plasmid – fast division causes high copy number, and vice versa.

Karyotype

In general, the karyotype is the characteristic chromosome complement of a eukaryote species. The preparation and study of karyotypes is part of cytogenetics.

Although the replication and transcription of DNA is highly standardized in eukaryotes, *the same cannot be said for their karyotypes*, which are often highly variable. There may be variation between species in chromosome number and in detailed organization. In some cases, there is significant variation within species. Often there is:

1. variation between the two sexes
2. variation between the germ-line and soma (between gametes and the rest of the body)
3. variation between members of a population, due to balanced genetic polymorphism
4. geographical variation between races
5. mosaics or otherwise abnormal individuals.

Also, variation in karyotype may occur during development from the fertilised egg.

The technique of determining the karyotype is usually called *karyotyping*. Cells can be locked part-way through division (in metaphase) in vitro (in a reaction vial) with colchicine. These cells are then stained, photographed, and arranged into a *karyogram*, with the set of chromosomes arranged, autosomes in order of length, and sex chromosomes.

Like many sexually reproducing species, humans have special gonosomes (sex chromosomes, in contrast to autosomes). These are XX in females and XY in males.

Historical Note

Investigation into the human karyotype took many years to settle the most basic question. How many chromosomes does a normal diploid human cell contain? In 1912, Hans von Winiwarter reported 47 chromosomes in spermatogonia and 48 in oogonia, concluding an XX/XO sex determination mechanism. Painter in 1922 was not certain whether the diploid number of man is 46 or 48, at first favouring 46. He revised his opinion later from 46 to 48, and he correctly insisted on man's having an XX/XY system.

New techniques were needed to definitively solve the problem:

1. Using cells in culture
2. Pretreating cells in a hypotonic solution, which swells them and spreads the chromosomes
3. Arresting mitosis in metaphase by a solution of colchicine
4. Squashing the preparation on the slide forcing the chromosomes into a single plane
5. Cutting up a photomicrograph and arranging the result into an indisputable karyogram.

It took until the mid-1950s for it to become generally accepted that the human karyotype include only 46 chromosomes. Considering the techniques of Winiwarter and Painter, their results were quite remarkable. Chimpanzees (the closest living relatives to modern humans) have 48 chromosomes.

Chromosomal Aberrations

Chromosomal aberrations are disruptions in the normal chromosomal content of a cell, and are a major cause of genetic conditions in humans, such as Down syndrome. Some chromosome abnormalities do not cause disease in carriers, such as translocations, or chromosomal inversions, although they may lead to a higher chance of birthing a child with a chromosome disorder. Abnormal numbers of chromosomes or chromosome sets, aneuploidy, may be lethal or give rise to genetic disorders. Genetic counseling is offered for families that may carry a chromosome rearrangement. The gain or loss of DNA from chromosomes can lead to a variety of genetic disorders. Human examples include:

- Cri du chat, which is caused by the deletion of part of the short arm of chromosome 5. "Cri du chat" means "cry of the cat" in French, and the condition was so-named because affected babies make high-pitched cries that sound like those of a cat. Affected individuals have wide-set eyes, a small head and jaw, moderate to severe mental health issues, and are very short.
- Down syndrome, usually is caused by an extra copy of chromosome 21 (trisomy 21). Characteristics include decreased muscle tone, stockier build, asymmetrical skull, slanting eyes and mild to moderate developmental disability.
- Edwards syndrome, which is the second-most-common trisomy; Down syndrome is the most common. It is a trisomy of chromosome 18. Symptoms include motor retardation, developmental disability and numerous congenital anomalies causing serious health problems. Ninety percent die in infancy; however, those that live past their first birthday usually are quite healthy thereafter. They have a characteristic clenched hands and overlapping fingers.
- Idic15, abbreviation for Isodicentric 15 on chromosome 15; also called the following names due to various researches, but they all mean the same; IDIC(15), Inverted duplication 15, extra Marker, Inv dup 15, partial tetrasomy 15
- Jacobsen syndrome, also called the terminal 11q deletion disorder. This is a very rare disorder. Those affected have normal intelligence or mild developmental disability, with poor expressive language skills. Most have a bleeding disorder called Paris-Trousseau syndrome.
- Klinefelter's syndrome (XXY). Men with Klinefelter syndrome are usually sterile, and tend to have longer arms and legs and to be taller than their peers. Boys with the syndrome are often shy and quiet, and have a higher incidence of speech delay and dyslexia. During puberty, without testosterone treatment, some of them may develop gynecomastia.
- Patau Syndrome, also called D-Syndrome or trisomy-13. Symptoms are somewhat similar to those of trisomy-18, but they do not have the characteristic hand shape.

- Small supernumerary marker chromosome. This means there is an extra, abnormal chromosome. Features depend on the origin of the extra genetic material. Cat-eye syndrome and isodicentric chromosome 15 syndrome (or Idic15) are both caused by a supernumerary marker chromosome, as is Pallister-Killian syndrome.
- Triple-X syndrome (XXX). XXX girls tend to be tall and thin. They have a higher incidence of dyslexia.
- Turner syndrome (X instead of XX or XY). In Turner syndrome, female sexual characteristics are present but underdeveloped. People with Turner syndrome often have a short stature, low hairline, abnormal eye features and bone development and a "caved-in" appearance to the chest.
- XYY syndrome. XYY boys are usually taller than their siblings. Like XXY boys and XXX girls, they are somewhat more likely to have learning difficulties.
- Wolf-Hirschhorn syndrome, which is caused by partial deletion of the short arm of chromosome 4. It is characterized by severe growth retardation and severe to profound mental health issues.

Chromosomal mutations produce changes in whole chromosomes (more than one gene) or in the number of chromosomes present.

- Deletion – loss of part of a chromosome
- Duplication – extra copies of a part of a chromosome
- Inversion – reverse the direction of a part of a chromosome
- Translocation – part of a chromosome breaks off and attaches to another chromosome.

Most mutations are neutral – have little or no effect. Chromosomal aberrations are the changes in the structure of chromosomes. It has a great role in evolution.

Human Chromosomes

Chromosomes can be divided into two types—autosomes, and sex chromosomes. Certain genetic traits are linked to your sex, and are passed on through the sex chromosomes. The autosomes contain the rest of the genetic hereditary information. All act in the same way during cell division.

Human cells have 23 pairs of large linear nuclear chromosomes, (22 pairs of autosomes and one pair of sex chromosomes) giving a total of 46 per cell. In addition to these, human cells have many hundreds of copies of the mitochondrial genome. Sequencing of the human genome has provided a great deal of information about each of the chromosomes.

Below is a table compiling statistics for the chromosomes, based on the Sanger Institute's human genome information in the Vertebrate Genome Annotation (VEGA) database. Number of genes is an estimate as it is in part based on gene predictions. Total chromosome length is an estimate as well, based on the estimated size of unsequenced heterochromatin regions.

Chromosome	*Genes*	*Total bases*	*Sequenced bases*
1	4,220	247,199,719	224,999,719
2	1,491	242,751,149	237,712,649
3	1,550	199,446,827	194,704,827
4	446	191,263,063	187,297,063
5	609	180,837,866	177,702,766
6	2,281	170,896,993	167,273,993
7	2,135	158,821,424	154,952,424
8	1,106	146,274,826	142,612,826
9	1,920	140,442,298	120,312,298
10	1,793	135,374,737	131,624,737
11	379	134,452,384	131,130,853
12	1,430	132,289,534	130,303,534
13	924	114,127,980	95,559,980
14	1,347	106,360,585	88,290,585
15	921	100,338,915	81,341,915
16	909	88,822,254	78,884,754
17	1,672	78,654,742	77,800,220
18	519	76,117,153	74,656,155
19	1,555	63,806,651	55,785,651
20	1,008	62,435,965	59,505,254
21	578	46,944,323	34,171,998
22	1,092	49,528,953	34,893,953
X (sex chromosome)	1,846	154,913,754	151,058,754
Y (sex chromosome)	454	57,741,652	25,121,652
Total	32,185	3,079,843,747	2,857,698,560

Chromosome Abnormality

A chromosome anomaly, abnormality or aberration reflects an atypical number of chromosomes or a structural abnormality in one or more chromosomes. A Karyotype refers to a full set of chromosomes from an individual which can be compared to a "normal" Karyotype for the species via genetic testing. A chromosome anomaly may be detected or confirmed in this manner. Chromosome anomalies usually occur when there is an error in cell division following meiosis or mitosis. There are many types of chromosome anomalies. They can be organized into two basic groups, numerical and structural anomalies.

Numerical Abnormalities

This is called Aneuploidy (an abnormal number of chromosomes), and occurs when an individual is missing either a chromosome from a pair (monosomy) or has more than two chromosomes of a pair (Trisomy, Tetrasomy, etc.). In humans, an example of a condition caused by a numerical anomaly is Down Syndrome, also known as Trisomy 21 (an individual with Down Syndrome has three copies of chromosome 21, rather than two). Turner Syndrome is an example of a monosomy where the individual is born with only one sex chromosome, an X. ''*euploidy it is of three types : monoploidy, diploidy, polyploidy. and this polyploidy is further divided into auto,endo and allopolyploid.*

Structural Abnormalities

When the chromosome's structure is altered. This can take several forms:

- Deletions: A portion of the chromosome is missing or deleted. Known disorders in humans include Wolf-Hirschhorn syndrome, which is caused by partial deletion of the short arm of chromosome 4; and Jacobsen syndrome, also called the terminal 11q deletion disorder.
- Duplications: A portion of the chromosome is duplicated, resulting in extra genetic material. Known human disorders include Charcot-Marie-Tooth disease type 1A which may be caused by duplication of the gene encoding peripheral myelin protein 22 (PMP22) on chromosome 17.
- Translocations: When a portion of one chromosome is transferred to another chromosome. There are two main types of translocations. In a reciprocal translocation, segments from

two different chromosomes have been exchanged. In a Robertsonian translocation, an entire chromosome has attached to another at the Centromere-in humans these only occur with chromosomes 13, 14, 15, 21 and 22.

- Inversions: A portion of the chromosome has broken off, turned upside down and reattached, therefore the genetic material is inverted.
- Rings: A portion of a chromosome has broken off and formed a circle or ring. This can happen with or without loss of genetic material.
- Isochromosome: Formed by the mirror image copy of a chromosome segment including the centromere.

Chromosome instability syndromes are a group of disorders characterized by chromosomal instability and breakage. They often lead to an increased tendency to develop certain types of malignancies.

Inheritance

Most chromosome anomalies occur as an accident in the egg or sperm, and are therefore not inherited. Therefore, the anomaly is present in every cell of the body.

Some anomalies, however, can happen after conception, resulting in Mosaicism (where some cells have the anomaly and some do not). Chromosome anomalies can be inherited from a parent or be "de novo".

This is why chromosome studies are often performed on parents when a child is found to have an anomaly.

Human sex refers to the processes by which an individual becomes either a male or female during development.

The Jost Paradigm

Under typical circumstances, the sex of an individual will be determined and expressed through the following mechanisms:

- Chromosomal Sex (genetic): Presence or absence of Y chromosome
- Gonadal Sex (Primary Sex Determination): Controlled by presence or absence of testis determining factor (TDF)
- Phenotypic Sex (Secondary Sex Differentiation): Determined by the hormonal products produced by the gonads.

Sex Determination

Sex Determination at the Chromosome Level

For the majority of individuals, sex determination is as simple as the presence or absence of a Y chromosome. Those individuals with a Y chromosome (including XXY, XXXY, etc.) will develop into males, and those without one will become female. Some individuals, however, will undergo what is referred to as primary sex reversal, whereby the X and Y chromosomes [cross over] and exchange genetic material. This relatively rare occurrence (approximately 1 in 20000 births) can lead to males with two X chromosomes and females with a Y chromosome.

Testis Determining Gene

During the late 1980s and early 1990s, coinciding with the mapping of the human genome, researchers began to look for the specific gene on the Y chromosome that, up until then, had been known as the testis determining factor (TDF). Through the study of individuals that underwent primary sex reversal (that is, XX males and XY females), researchers determined that the TDF must lie on the Y chromosome in a location that would permit its exchange to the X chromosome during cross over. In 1985, Dr. David C. Page published an article in Nature boldly stating that the TDF was the ZFY gene on the Y chromosome. However, Dr. MS Palmer later discovered a ZFY analogue on the X chromosome, providing evidence that ZFY was in fact not the TDF. Eventually, Dr. Peter Koopman was able to prove that the SRY gene is the TDF from studies on XX males.. The following evidence further supports this claim:

- SRY is Y specific, and there is no analogue on the X chromosome.
- SRY is deleted or mutated in XY females
- It undergoes expression within the testis at the time of testis differentiation.
- Its sequence suggests that its protein has a DNA binding motif because it has high homology to an 80 amino acid long DNA binding region (HMG box).

SRY a Repressor?

Recently, it has been suggested by some that the SRY gene acts as a repressor or inhibitor of another gene, "Z", that is involved in female development. Previously, it was stated that the SRY sequence

- Women have a larger hip section than men, an adaptation for giving birth to infants with large skulls.

Skeleton and Muscular System

Strength, Power and Muscle Mass

On average, males are stronger than females. This is due to females, on average, having less total muscle mass than males. Females also have lower muscle mass in comparison to total body weight. Gross measures of upper body strength suggest an average 40-50% difference between the sexes, compared to a 30% difference in lower body strength. One study of muscle strength in the elbows and knees—in 45 and older males and females—found the strength of females to range from 42 to 63% of male strength. Males are not stronger due to greater strength of individual muscle fibres, but due to more fibres: a greater total muscle mass. The greater muscle mass of males is in turn due to a greater capacity for muscular hypertrophy as a result of men's higher levels of testosterone. Males remain stronger than females, when adjusting for differences in total body weight. This is due to the higher male muscle-mass to body-weight ratio. The uterus—as it is a part of the female reproductive system—is a muscle found only in females. The uterus may be the single strongest muscle in comparison to body weight across both sexes.

Skeleton

- In men, the second digit (index finger) tends to be shorter than the fourth digit (ring finger), while in women the second digit tends to be longer than the fourth.
- Men have a more pronounced 'Adam's Apple' or thyroid cartilage due to larger vocal cords (and deeper voices).
- On average, men have longer canine teeth than women.
- Female skulls and head bones have a different shape than male skulls. One difference is in the roundness of the eye cavities, another difference is the shape of the jaw.
- Male and female pelvises are shaped differently. The female pelvis features a wider pelvic cavity, which is necessary when giving birth. The female pelvis has evolved to its maximum width for childbirth — an even wider pelvis would make women unable to walk. In contrast, human male pelves are not constrained by the need to give birth and are therefore more optimized for walking. As a result, the female pelvis is larger

and broader than the male pelvis which is taller, narrower, and more compact. The female inlet is larger and oval in shape, while the male inlet is more heart-shaped.

- However, contrary to popular belief, males and females do not differ in the number of ribs; both have twelve pairs.

Comparative and social psychologists have observed that males and females, in general, differ in the way they carry books while walking. Upon using a classification system of the five common methods of carrying books, a high percentage of females will partially cover their body with the books they are carrying, such as by holding them in front of the chest. Most males carry their books at the side of the body, leaving the front uncovered. The most common explanation of this observation is that women typically have less strength than men, making it difficult to balance, and resulting in the need to rest the objects they are carrying on their bodies. Some psychologists hypothesize that it is a maternal instinct in many women causing them to carry inanimate objects in a protective manner.

Skin and Hair

Skin

Male skin is thicker (more collagen) and oilier (more sebum) than female skin.

The skin of females is warmer on average than that of males.

Hair

On average, males have more body hair than women. Males have relatively more of the type of hair called terminal hair, especially on the face, chest, abdomen and back. In contrast, females have more vellus hair. Vellus hairs are smaller and therefore less visible.

Baldness is much more common in males than in females. The main cause for this is *male pattern baldness* or androgenic alopecia. Male pattern baldness is a condition where hair starts to get lost in a typical pattern of receding hairline and hair thinning on the crown, and is caused by hormones and genetic predisposition.

Colour

On average and after the end of puberty, males may have darker skin than females as well as darker hair. Male eyes are also perhaps more likely to be one of the darker eye colours. Conversely, women are possibly lighter-skinned than men in most or all populations. The differences in colour are mainly caused by higher levels of melanin in the skin, hair and eyes in males. In one study, almost twice as many females as males had red or auburn hair. A higher proportion of females were also found to have blond hair, whereas males were more likely to have black or dark brown hair.

Another study found green eyes, which are a result of lower melanin levels, to be much more common in women than in men, at least by a factor of two. However, the most recent and detailed study on the matter sampled a relatively large amount of subjects in Poland and found that while women indeed did tend to have a lower frequency of black hair and black eyes then men, men on the other hand had a higher frequency of such features such as red hair, blond hair, blue eyes and light skin then women did with the most striking feature being the significantly higher frequency of reddish blond hair in men then women. This phenomenon may be correlated with the theory that women have a higher frequency of genetic recombination then men and therefore tend to have less phenotypical variation then men in any given population This theory was previously supported by a study that stated that There are no significant differences in eye-colour distributions between females and males.

Sexual Organs and Reproductive Systems

Men and women have different sex organs. Women have two ovaries that produce eggs, and uterus which is connected to a vagina. Men have testicles that produce sperm. The testicles are placed in the scrotum behind the penis. The male penis and scrotum are external extremities, whereas the female sex organs are placed “inside” the body.

Reproductive Capacity and Cost

Men typically produce billions of sperm each month, many of which are capable of fertilization. Women typically produce one egg a month that can be fertilized into an embryo. Thus during a lifetime men are able to father a significantly greater number of children than women can give birth to. The most fertile woman, according to the Guinness Book of World Records, was the wife of Feodor Vassilyev of Russia (1707–1782) who had 67 surviving children. The most prolific father of all time is believed to be the last Sharifian Emperor of Morocco, Mulai Ismail (1646–1727) who reportedly fathered more than 800 children from a harem of 500 women.

Fertility

Female fertility declines after age 30 and ends with the menopause. Pregnancy in the 40s or later has been correlated with increased chance of Down's Syndrome in the children. Men are capable of fathering children into old age. Paternal age effects in the children include multiple sclerosis, autism, breast cancer and schizophrenia, as well as reduced intelligence. Adriana Iliescu was reported as the world's oldest woman to give birth, at age 66. Her record stood until Maria del Carmen Bousada de Lara gave birth to twin sons at Sant Pau Hospital in Barcelona, Spain on December 29, 2006, at the age of 67. In both cases IVF was used. The oldest known father was former Australian miner Les Colley, who fathered a child at age 93.

Brain and Nervous System

Brain

Males have larger brains than females. On average, male brains have approximately 4 % more cells and weigh 100 grams more than female brains do. However, both sexes have a similar brain weight to body weight ratio. Female brains are more compact than male brains in that, though smaller, they are more densely packed with neurons, particularly in the region responsible for language.

There are also differences in the structure of and in specific areas of the brain. Men have larger left inferior parietal lobes, while women have larger Wernicke's and Broca's areas. Evidence of gender differences in the size of the corpus callosum is ambiguous. Also, females may have their language functions more evenly distributed in both cerebral hemispheres, while in males they are more concentrated in the left hemisphere. This puts males more at risk for language disorders like dyslexia.

It has been argued that the Y chromosome is primarily responsible for males being more susceptible to mental illnesses. Women generally have faster blood flow to their brains and lose less brain tissue as they age than men do. Depression and chronic anxiety are much more common in women than in men, (it has been speculated, by some, that this is due to difference in the brain's serotonin system).

Sensory Systems

- Females have a more sensitive sense of smell than males, both in the differentiation of odors, and in the detection of slight or faint odors.

- There is also indication that females are better at discerning differences in colours, while males are more aware of, and capable of discerning movement.
- Females have more pain receptors in the skin. That may contribute to the lower pain tolerance of women.
- Males have stronger spatial analysis capabilities, in both navigation and awareness.

Tissues and Hormones

- Women generally have a higher body fat percentage than men.
- Women usually have lower blood pressure than men, and women's hearts beat faster, even when they are asleep.
- Men generally have more muscle tissue mass, particularly in the upper body.
- Men and women have different levels of certain hormones. Men have a higher concentration of androgens while women have a higher concentration of estrogens. The main male-associated hormone is testosterone.
- Adult men have approximately 5.2 million red blood cells per cubic millimeter of blood, whereas women have approximately 4.6 million.

Health

Life Span

Women live longer than men in most countries. One possible explanation is that more men die young because of war, criminal activity, and accidents. The gap between males and females is decreasing in many developed countries as more women take up unhealthy practices that were once considered masculine like smoking and drinking alcohol, and more men practice healthier living. In Russia, however, the sex-associated gap has been increasing as male life expectancy declines.

Health Issues

The World Health Organization (WHO) has produced a number of reports on gender and health. The following trends are shown:

- Overall rates of mental illness are similar for men and women. There is no significant gender difference in rates of schizophrenia and bipolar depression. Women are more likely to suffer from unipolar depression, anxiety, eating disorders, and post-traumatic stress disorder. Men are more likely to suffer from alcoholism and antisocial personality disorder.

- Worldwide, more men than women are infected with HIV. The exception is sub-Saharan Africa, where more women than men are infected.
- Adult males are more likely to be diagnosed with tuberculosis.
- Before menopause, women are less likely to suffer from cardiovascular disease. However, after age 60, the risk for both men and women is the same.
- Overall, men are more likely to suffer from cancer, with much of this driven by lung cancer. In most countries, more men than women smoke, although this gap is narrowing especially among young women.
- Women are twice as likely to be blind as men. In developed countries, this may be linked to higher life expectancy and age-related conditions. In developing countries, women are less likely to get timely treatments for conditions that lead to blindness such as cataracts and trachoma.
- Women are more likely to suffer from osteoarthritis and osteoporosis.

Anterior cruciate ligament injuries, especially in basketball, occur more often in women than in men.

Certain conditions are X-linked recessive, in that the gene is carried on the X chromosome. Genetic females (XX) will show symptoms of the disease only if both their X chromosomes are defective with a similar deficiency, whereas genetic males (XY) will show symptoms of the disease if their only X chromosome is defective. (A woman may carry such a disease on one X chromosome but not show symptoms if the other X chromosome works sufficiently.)

For this reason, such conditions are far more common in males than in females. Examples of X-linked recessive conditions are colour blindness, hemophilia, and Duchenne muscular dystrophy. From conception to death, but particularly before adulthood, females are less vulnerable than males to developmental difficulties and chronic illnesses. This could be due to females having two x chromosomes instead of just one, or in the reduced exposure to testosterone.

Testosterone

In the 1930s, Alfred Jost determined that the presence of testosterone was required for Wolffian duct development in the female rabbit.

Müllerian Inhibiting Substance

Jost also observed that while testosterone was required for Wolffian duct development, the regression of the Müllerian duct was due to another substance. This was later determined to be Müllerian inhibiting substance (MIS), a 140 kD dimeric glycoprotein that is produced by sertoli cells. MIS blocks the development of Müllerian ducts, promoting their regression.

5-alpha Dihydrotestosterone (DHT)

Testosterone is converted to the more potent DHT by 5-alpha reductase. DHT is necessary to exert androgenic effects farther from the site of testosterone production, where the concentrations of testosterone are too low to have any potency.

A 5-alpha reductase deficiency results in an androgen disorder characterized by female phenotype or severely undervirilized male phenotype with development of the epididymis, vas deferens, seminal vesicle, and ejaculatory duct, but also a pseudovagina.

Pathologies

The following disorders are caused by a malfunction in the sex determination and differentiation process:

- Congenital Adrenal Hyperplasia-Inability of adrenal to produce sufficient cortisol, leading to increased production of testosterone resulting in severe masculinization of 46 XX females.
- Persistent Müllerian Duct Syndrome-A rare type of pseudohermaphroditism that occurs in 46 XY males, caused by either a mutation in the Müllerian inhibiting substance (MIS) gene, on 19p13, or its type II receptor, 12q13. Results in a retention of Müllerian ducts (persistance of rudimentary uterus and fallopian tubes in otherwise normally virilized males), unilateral or bilateral undescended testes and sometimes causes infertility.
- Male Pseudohermaphroditism-Failure of androgen production or inadequate androgen response, which can cause incomplete masculinization in XY males. Varies from mild failure of

masculinization with undescended testes to complete sex reversal and female phenotype (Androgen insensitivity syndrome)

- Swyer syndrome. A form of complete gonadal dysgenesis, mostly due to mutations in the first step of sex determination; the SRY genes.

Sex-determination System

A sex-determination system is a biological system that determines the development of sexual characteristics in an organism. Most sexual organisms have two sexes. In many cases, sex determination is genetic: males and females have different alleles or even different genes that specify their sexual morphology. In animals, this is often accompanied by chromosomal differences.

In other cases, sex is determined by environmental variables (such as temperature) or social variables (the size of an organism relative to other members of its population). The details of some sex-determination systems are not yet fully understood.

Bibliography

Aldridge, S.: *The Thread of Life: The Story of Genes and Genetic Engineering*, Cambridge University Press, Cambridge, 1996.

Allan, V., B. Backley, L. Felperin, N. James and H. Gee: *Sight and Sound Supplement*, BFI, London, 1996.

Astor, G.: *The "Last" Nazi: The Life and Times of Dr. Joseph Mengele*, Weidenfeld and Nicolson, London, 1985.

Ayala, F. J.: *The Genetic Structure of Populations*, W.H. Freeman & Co., San Francisco, California, 1977.

Bajema, C. J.: *Natural Selection in Human Populations, the Measurement of Ongoing Genetic Evolution in Contemporary Societies*, Wiley, New York, 1971.

Barnaby, W.: *Plague Makers: The Secret World of Biological Warfare*, Vision Paperbacks, London, 1999.

Bast, R.C. Jr.: *Cancer Medicine*, Decker, Hamilton, 2000.

Bauer, M.: *Resistance to New Technology: Nuclear Power, Information Technology and Biotechnology*, Cambridge University Press, Cambridge, England, 1995.

Berg, Paul, and Singer, Maxine: *Dealing with Genes: The Language of Heredity. Mill Valley*, University Science Books, CA, 1992.

Bernard, C.: *An Introduction to the Study of Experimental Medicine*, Dover Publications, New York, 1957.

Berry, R. J.: *Inheritance and Natural History*: Collins, London, 1977.

Bradley, W.G.: *Neurology in Clinical Practice*, Butterworth Heinemann, Boston, 2000.

Branagh, K., S. Lady, and F. Darabont: *Mary Shelley's Frankenstein: The Classic Tale of Terror Reborn*, Newmarket Press, London, 1994.

Brookes, Martin: *Get a Grip on Genetics, Time Life Books*, East Sussex, England, 1998.

Brosnan, J.: *The Primal Scream: A History of Science Fiction Film*, Orbit Books, London, 1991.

Bukatman, S.: *Blade Runner*, BFI, London, 1997.

Campbell, Neil A.; Brad Williamson; Robin J. Heyden: *Biology: Exploring Life*, Pearson Prentice Hall, Boston, Massachusetts, 2006.

Crick, F.: *Life Itself: Its Origin and Nature*, W.W. Norton, New York, 1982.

Curtis, M.: *The Geometry of DNA: A Structural Revision*, Blue Gallery, London, 1996.

Dams, R.D.: *Principles of Neurology*, McGraw Hill, New York, 1997.

Dawkins, R.: *The Selfish Gene*, Oxford University Press, New York, 1976.

Diaz, E.: *Microbial Biodegradation: Genomics and Molecular Biology,* Caister Academic Press, UK, 2008.

Dobzhanshy, T.: *Genetics of the Evolutionary Process*, Columbia University Press, New York, 1970.

Dulbecco, R.: *The Design of Life*, Yale University Press, New Haven, Connecticut, 1987.

Dunbar, Robert E.: *Heredity*, Franklin Watts Publisher, New York, 1978.

Edey, M. A., and Johanson, D. C.: *Blueprints: Solving the Mystery of Evolution*, Little, Brown and Co., Boston, Mass, 1989.

Fisher, R. A.: *The Genetical Theory of Natural Selection*, Clarendon Press, Oxford, 1930.

Fritz, A.: *International Classification of Diseases for Oncology*, World Health Organization, Geneva, 2000.

Gall, Joseph G.: *Landmark Papers in Cell Biology,* Cold Spring Harbor Laboratory Press, Plainview, NY, 2001.

Garza-Valdes, L.A.: *The DNA of God?* Hodder and Stoughton, London, 1998.

George S. Paul: *Beyond Humanity: Cyber Evolution and Future Minds*, Charles River Media, Roackland, 1996.

Gillis, Justin: *Drug Firms, Gene Labs to Map Genetic Code*, The Daily News, Longview, WA, 1999.

Glover, D. M., and Hames, B. D.: *Genes and Embryos*, Oxford University Press, New York, 1989.

Goldman, L.; Ausiello, D. A.: *Cecil Textbook of Medicine*, Saunders, Philadelphia, 2004.

Griffiths A.J.F.: *Introduction to Genetic Analysis*, W.H. Freeman and Company, New York, USA, 2005.

Halacy, D.S., Jr.: *Genetic Revolution, Shaping Life for Tomorrow,* Harper & Row, Publishers, New York, 1974.

Hamerton, J. L.: *Human Cytogenetics*, Academic Press, New York, 1971.

Hanley, R.: *Is Data Human? The Metaphysics of Star Trek*, Boxtree, London, 1998.

Hartl, Daniel L.: *Basic Genetics*, Jones and Bartlett Publishers, Boston, 1991.

Haubrich, W.S.: *Bockus Gastroenterology*, Saunders, Philadelphia, 1995.

Heider, J. and Rabus, R: *Microbial Biodegradation: Genomics and Molecular Biology*, Caister Academic Press, UK, 2008.

Holtz, Robert D.; William, Kovacs D.: *An Introduction to Geotechnical Engineering*, Prentice Hall, UK, 1981.

Jameson, J. L.: *Principles of Molecular Medicine*, Humana Press, Totowa, 1998.

Jonoska, N.: *Self-Assembling DNA Graphs, DNA-Based Computers VIII*, Springer-Verlag, Berlin, 2003.

Kelves, D.: *In the Name of Eugenics: Genetics and the Uses of Human Heredity*. Harmondsworth: Penguin.

Klein, Aaron E.: *Threads of Life: Genetics from Aristotle to DNA*, The Natural History Press, Garden City, New York, 1955.

Klug, William S. and Michael R.: Cummings. *Essentials of Genetics*, Prentice Hall, New Jersey, 1996.

Lakoff, G. and M. Johnson: *Metaphors We Live By*, University of Chicago Press, Chicago, 1980.

Lewin B.: *Genes VII*, Oxford University Press Inc., New York, USA, 2000.

Lewontin, R.C.: *The Doctrine of DNA: Biology as Ideology*, Penguin, Harmondsworth, 1993.

Lyon, Jeff and Gorner, Peter. Altered Fates: *Gene Therapy and the Retooling of Human Life*, W. W. Norton and Company, New York, 1995.

Margulis, L.: *Symbiosis in Cell Evolution*, W.H. Freeman, San Francisco, 1981.

Mayr, E.: *Change of Genetic Environment and Evolution*, Allen and Unwin, London, 1954.

Mayr, Ernst: *The Growth of Biological Thought: Diversity, Evolution, and Inheritance*, Harvard University Press, Cambridge, 2000.

Migloni, G.S.: *Dictionary of Plant Genetics and Molecular Biology*, Hawthorne Press, New York, 1998.

Nelkin, D. and M.S. Lindee: *The DNA Mystique: The Gene as a Cultural Icon*, W.H. Freeman, New York, 1995.

Nottingham, S.F.: *Eat Your Genes: How Genetically Modified Food is Entering Our Diet*, Zed Books, London, 1998.

Rietman, Ed.: *Molecular Engineering of Nanosystems*, Springer, New York, 2001.

Schummer, J.: *Interdisciplinary Issues in Nanoscale Research*, IOS Press, Amsterdam, 2004.

Scriver, C.R.: *The Metabolic and Molecular Basis of Inherited Disease*, McGraw Hill, New York, 2001.

Stebbins, G. L.: *Darwin to DNA, Molecules to Humanity*, W. H. Freeman, San Francisco, 1982.

Sturtevant, Alfred: *History of Genetics*, Harper and Row, New York, 1965.

Stwertka, Eve and Albert: *Genetic Engineering*, Franklin Watts, New York, 1982.

Suzuki, D., and Knudtson, P.: *Genethics: The Clash Between the New Genetics and Human Values*, Harvard University Press, Cambridge, Mass, 1989.

Watson, J. D.: *The Double Helix*, Antheneum, New York, 1968.

Williams, J.G. and R.K. Patient: *Genetic Engineering*, IRL Press, Oxford, England, 1988.

Wilson, E. O., and Lumden, C.: *Genes, Mind, and Culture: The Evolutionary Process*, Harvard University Press, Cambridge, Mass, 1981.

Wilson, E. O.: *Biophilia*, Harvard University Press, Cambridge, Mass, 1985.

Winchester, A. M.: *Heredity, Evolution and Humankind*, West Publishing Co., St. Paul, Minn., 1976.

Winston, R.: *The Future of Genetic Manipulation*, Phoenix, London, 1997.

Wright, L.: *Twins: Genes, Environment and the Mystery of Human Identity*, Weidenfeld & Nicolson, London, 1997.

Zimmerman, E. G.: *Karyology, Systematics, and Chromosomal Evolution in the Rodent Genus, Sigmodon*, Michigan State Univ., UK 1970.

Index